Advances in Intelligent Systems and Computing

Volume 598

About this Series

The series "Advances in Intelligent Systems and Computing" contains publications on theory, applications, and design methods of Intelligent Systems and Intelligent Computing. Virtually all disciplines such as engineering, natural sciences, computer and information science, ICT, economics, business, e-commerce, environment, healthcare, life science are covered. The list of topics spans all the areas of modern intelligent systems and computing.

The publications within "Advances in Intelligent Systems and Computing" are primarily textbooks and proceedings of important conferences, symposia and congresses. They cover significant recent developments in the field, both of a foundational and applicable character. An important characteristic feature of the series is the short publication time and world-wide distribution. This permits a rapid and broad dissemination of research results.

More information about this series at http://www.springer.com/series/11156

Tareq Ahram · Waldemar Karwowski
Editors

Advances in Human Factors, Software, and Systems Engineering

Proceedings of the AHFE 2017 International Conference on Human Factors, Software, and Systems Engineering, July 17–21, 2017, The Westin Bonaventure Hotel, Los Angeles, California, USA

Springer

Editors
Tareq Ahram
University of Central Florida
Orlando, FL
USA

Waldemar Karwowski
University of Central Florida
Winter Park, FL
USA

ISSN 2194-5357 ISSN 2194-5365 (electronic)
Advances in Intelligent Systems and Computing
ISBN 978-3-319-60010-9 ISBN 978-3-319-60011-6 (eBook)
DOI 10.1007/978-3-319-60011-6

Library of Congress Control Number: 2017943065

Printed on acid-free paper

This Springer imprint is published by Springer Nature
The registered company is Springer International Publishing AG
The registered company address is: Gewerbestrasse 11, 6330 Cham, Switzerland

Advances in Human Factors and Ergonomics 2017

AHFE International Applied Human Factors and Ergonomics Conference 2017 www.ahfe.org

AHFE 2017 Series Editors

Tareq Z. Ahram, Florida, USA
Waldemar Karwowski, Florida, USA

8th International Conference on Applied Human Factors and Ergonomics and the Affiliated Conferences

Proceedings of the AHFE 2017 International Conference on Human Factors, Software, and Systems Engineering, July 17–21, 2017, The Westin Bonaventure Hotel, Los Angeles, California, USA

Advances in Affective and Pleasurable Design	*WonJoon Chung and Cliff (Sungsoo) Shin*
Advances in Neuroergonomics and Cognitive Engineering	*Carryl Baldwin*
Advances in Design for Inclusion	*Giuseppe Di Bucchianico and Pete Kercher*
Advances in Ergonomics in Design	*Francisco Rebelo and Marcelo Soares*
Advances in Human Error, Reliability, Resilience, and Performance	*Ronald L. Boring*
Advances in Human Factors and Ergonomics in Healthcare and Medical Devices	*Vincent G. Duffy and Nancy Lightner*
Advances in Human Factors in Simulation and Modeling	*Daniel N. Cassenti*
Advances in Human Factors and System Interactions	*Isabel L. Nunes*
Advances in Human Factors in Cybersecurity	*Denise Nicholson*
Advances in Human Factors, Business Management and Leadership	*Jussi Kantola, Tibor Barath and Salman Nazir*
Advances in Human Factors in Robots and Unmanned Systems	*Jessie Chen*
Advances in Human Factors in Training, Education, and Learning Sciences	*Terence Andre*
Advances in Human Aspects of Transportation	*Neville A. Stanton*

(continued)

(continued)

Advances in Human Factors, Software, and Systems Engineering	*Tareq Z. Ahram and Waldemar Karwowski*
Advances in Human Factors in Energy: Oil, Gas, Nuclear and Electric Power Industries	*Paul Fechtelkotter and Michael Legatt*
Advances in Human Factors, Sustainable Urban Planning and Infrastructure	*Jerzy Charytonowicz*
Advances in the Human Side of Service Engineering	*Louis E. Freund and Wojciech Cellary*
Advances in Physical Ergonomics and Human Factors	*Ravindra Goonetilleke and Waldemar Karwowski*
Advances in Human Factors in Sports, Injury Prevention and Outdoor Recreation	*Tareq Z. Ahram*
Advances in Safety Management and Human Factors	*Pedro Arezes*
Advances in Social & Occupational Ergonomics	*Richard Goossens*
Advances in Ergonomics of Manufacturing: Managing the Enterprise of the Future	*Stefan Trzcielinski*
Advances in Usability and User Experience	*Tareq Ahram and Christianne Falcão*
Advances in Human Factors in Wearable Technologies and Game Design	*Tareq Ahram and Christianne Falcão*
Advances in Communication of Design	*Amic G. Ho*
Advances in Cross-Cultural Decision Making	*Mark Hoffman*

Preface

The discipline of Human Factors, Software, and Systems Engineering provides a platform for addressing challenges in human factors, software, and systems engineering that both pushes the boundaries of current research and responds to new challenges, fostering new research ideas. In this book, researchers, professional software & systems engineers, human factors and human systems integration experts from around the world addressed societal challenges and next-generation systems and applications for meeting them. The books address topics from evolutionary and complex systems, human systems integration to smart grid and infrastructure, workforce training requirements, systems engineering education, and even defense and aerospace. It is sure to be one of the most informative systems engineering events of the year.

This book focuses on the advances in the Human Factors, Software, and Systems Engineering, which are a critical aspect in the design of any human-centered technological system. The ideas and practical solutions described in this book are the outcome of dedicated research by academics and practitioners aiming to advance theory and practice in this dynamic and all-encompassing discipline. This book is divided into two main sections:

I. Methods and Techniques
II. Applications

Each section contains research paper that has been reviewed by members of the International Editorial Board. Our sincere thanks and appreciation to the Board members as listed below:

P. Fechtelkotter, USA
F. Fischer, Brazil
S. Fukuzumi, Japan
R. Goonetilleke, Hong Kong
C. Grecco, Brazil
N. Jochems, Germany
G.J. Lim, USA
D. Long, USA
R. Martins, Brazil
M. Mochimaru, Japan
C. O'Connor, USA
C. Orłowski, Poland
H. Parsaei, Qatar
S. Pickl, Germany
S. Ramakrishnan, USA
Jose San Martin Lopez, Spain
K. Santarek, Poland
M. Shahir Liew, Malaysia
D. Speight, UK
M. Stenkilde, Sweden
T. Winkler, Poland
H. Woodcock, UK

We hope that this book, which is the international state of the art in human factors in software and systems engineering, will be a valuable source of theoretical and applied knowledge enabling human-centered design of variety of products, services, and systems for global markets.

July 2017

Tareq Z. Ahram
Waldemar Karwowski

Contents

Methods and Techniques

Usage Perspective Development Approach in the Fuzzy Front End

Sigmund Schimanski(✉)

Faculty 6 – Electrical, Information and Media Engineering, Automation/Computer Science, Working Group: Human Factors Engineering, University of Wuppertal, Rainer-Gruenter-Str. 21, 42119 Wuppertal, Germany
schimanski@uni-wuppertal.de

Abstract. The development of complex systems holds various networking and combination opportunities with a corresponding number of desired and unwanted effects on system functions and/or human-machine interactions. Their vast number makes these effects difficult to control during development. The usage-perspective based approach we try to develop is intended to provide the primary starting point for deriving the necessary requirements for finding solutions for technical problems or realising existing ideas. The immersion into usage processes promises a reliable and holistic derivation of requirements for beneficial product solutions in early development phases. The multi-usage perspective approach will be made available to relevant innovation engineers in order to broaden their own perspective experience. It allows the analysis and synthesizing of specific structurated solution concepts (Innovateurship) which shall leads to implementable products in further steps of the development. This approach increases the appeal of the product as well as the productivity of the provider.

Keywords: Usage perspective · Innovateur · Innovateurship · Innovatorship · Perspective engineering · Human factors · Ergonomics · Usability · Requirements elicitation · Design methodologies · Systems engineering · HCI · HMI · HSI · HFI

1 Introduction

The digitization and networking of complex products, systems and services (hereafter referred to as products) offer competitive advantages to companies – but on the other hand, they also present challenges of obscure production dependencies, a low level of market acceptance, and the loss of control. In value creation processes, many beneficial potentials are only detected in practice, and can't be integrated because they can cause unpredictable interactions and thus system interruptions. As a result, developers focus on the functional implementation of solutions and thus often only on the task without considering the adjacent system conditions. There is hardly any empirical knowledge about the success criteria in the development of sustainable innovations, particularly about their effects on usability and the long-term perceived quality. No known model considers the systematic and primary application of a structured usage perspective in

T. Ahram and W. Karwowski (eds.), *Advances in Human Factors, Software, and Systems Engineering*, Advances in Intelligent Systems and Computing 598,
DOI 10.1007/978-3-319-60011-6_1

the early innovation development phases. It can be assumed that a focus on the use and a corresponding categorization of the usage phases can help to make the determination of the initial requirements management more valid, and thus create a basis for the development of usable human-machine systems [1]. For sustainable decision-making abilities in processes of innovation developments, we will investigate the attainment of sufficient situation awareness [2] and criteria of role immersion in the following.

2 Purpose

This paper assumes that the development has to be usage-centred and perspective-based in order to achieve a broad product acceptance these days, and thus increase the economic success. Thus, the usage and role focussed development approach of interdisciplinary teams enables them to immerse into relevant value-added perspectives in order to sensitise to an ambivalent system perspective already in early phases [3]. The model allows them to work together as well as individually on a development issue, and to get the current status from the relevant perspectives at any time.

The project investigates a correspondingly granulated, human-oriented usage perspective to derive a usage-oriented procedure model, and thus to extend the classical value chain or product life-cycle [4], in order to sufficiently specify requirements in value phases for the various usage conditions in the context of digitization. We aim to create a criteria catalogue of the product use chain that can be integrated into the product life-cycle, and can be applied in early development phases. By deriving the usage aspect, we want to create a standardized approach that makes it possible to achieve a reliable procedure in early development stages in the context of innovation management as well as sustainable, attractive and valuable product and process solutions. This makes it possible to contrast the requirements derived from the usage perspective with the planned functions, and thus verify the reliability of solution approaches in the context of human-machine interaction and estimate the resulting subsequent effort. The model considers criteria of usage phases as well as a perspective-considering synthesis process. Further steps beyond the early phases are not part of this investigation. This paper also presupposes the existence of an invention and specific problem definition – and thus addresses neither.

3 Methodology

This paper is based on the elaboration and evaluation of the usage model with interviews of mobility users. First, we regarded the classical product life-cycle model [5] to construct the model. Typical elements of this cycle are, for example, the development at the beginning and the decline at the end. An element in between is the introduction to the market provided by distribution or service. The center point of the concept is the practice-granulated usage (Fig. 1), which forms the primary influence on the early development stages.

The approach consists of various actors and/or content areas. They consist of various aspects of the product's life that are defined by the interaction, i.e. the usage,

Fig. 1. Usage phases in the product-life cycle [1]

with the user. The user represents the consumer, and thus the key element of the product's life. The user's behaviour, habits and requirements should make the development of new, usage-centred products possible.

With the aim to develop a general model applicable to different kinds of products, we generalised the gained usage phases in the context of electric mobility, and examined them in an open innovation survey in the context of interconnected and intermodal mobility. The findings and ideas were later presented to a focus group of engineer students. The young, potential developers were asked to examine and optimise the feasibility of the results of the open innovation survey.

The aim was to validate or falsify the assumed influence factors. An additional intention was to discover further relevant areas and the implicit knowledge [6] of the users. To achieve this, the survey included unspecific questions, which allowed the users to freely share aspects that were relevant for them (e.g. open questions regarding the interconnectivity between electric mobility and living) (Fig. 2).

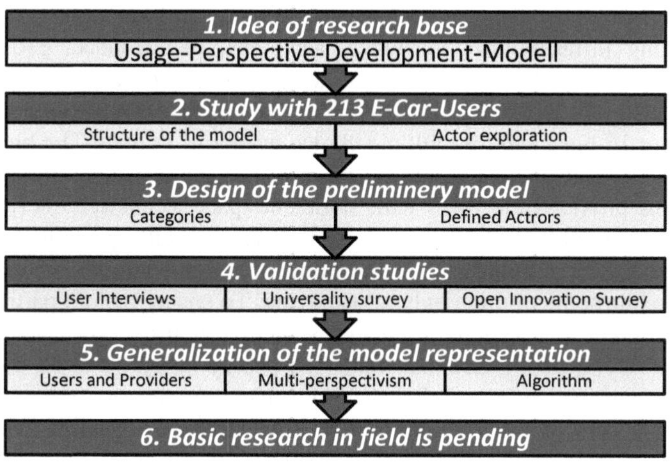

Fig. 2. Study set-up for the investigation of the usage-perspective model

The aim of the open innovation survey claimed was to collect opinions, criticism and "innovative", i.e. new, ideas about mobility. By not using specific questions, the users (motorists as well as users of public transport) can freely express their thoughts and wishes about various keywords in the context of individual transport, public transport and their interconnectivity. The survey provided text fields for free answers and gave only basic impulses for the answers in order to determine the aspects that were relevant for the users themselves. Finally, the moderated focus group was used to evaluate whether the aspiring engineers were able to use their knowledge to assess the results.

4 The Electric Vehicle Survey

After the determination of the current state of research, an empirical study with 213 interview partners and further surveys, the usage phases were hypothetically defined with a deductive-nomological model, and subsequently validated in usage-oriented development projects and focus groups. After the validation stage, phases, such as buying, production, insurance, maintenance, use etc., became purchase, protection, regulations, social and technical infrastructure, application, maintenance, updates etc. The derived structured usage processes allows the consideration of relevant system use scenarios in the early development phases. As an additional result, the requirement catalogue becomes more consistent and has to be revised less often. Thus, we could identify the resulting beneficial potentials of a system solution, and specify them for further steps (laboratory and field tests).

4.1 Structure and Conduction

The survey on electric mobility had eleven questions about various topics: general personal information, general vehicle details, purchase information, act of purchase, usage, charging infrastructure, insurance, accidents, maintenance, exploitation and concluding comments. The topics were asked in random order to avoid a sequence effect.

There were 16 open and 69 multiple-choice questions in total. The high amount of open questions is a distinctive feature of this survey. The text fields for answering the open questions provided the users with sufficient space to answer freely and in their own words. The aim of this form of inquiry is to register all aspects that matter to the users. Some of the 69 multiple-choice questions offered the option to choose more than one answer, while some of them only allowed one. To ensure that no participant would feel excluded and be able to find a suitable answer option, the survey often provided the option "other" along with a free text field. The survey lasted 30 to 60 min (Fig. 3).

Following the evaluation of the electric vehicle survey, we adjusted the usage model. The answers of the 213 participants generated new criteria, which took all relevant influence factors in the particular context of electric mobility into consideration.

4.2 Participants of the Electric Vehicle Survey

The survey was designed for electric vehicle users and owners. To gather as many opinions as possible, the users could either choose between an online and a telephone

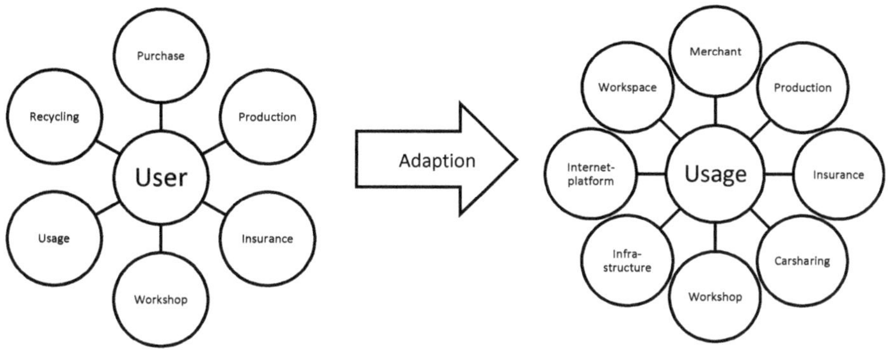

Fig. 3. Development of phases in the usage model [1]

survey. 213 users participated in total over a period of three months – 170 participants used the online and 43 participants the telephone survey.

4.3 Evaluation of the Survey with Regard to the Usage Model

The result of the user survey was a user-centred behaviour, usage, requirement and risk analysis, which listed potential consumer preferences, user profiles, needs, errors, defects and accidents as well as their probability of occurrence, meaning, and probability of detection. First, it can be noted that the survey confirmed and supported all previously assumed influence factors of the usage model.

An entire section of the survey, with eleven questions in total, was about the *charging infrastructure.* This section represents the area of usage in the assumed usage model as the use of vehicles refers directly to the infrastructure. The frequency of replies as well as the answers themselves justifies an inclusion of the infrastructure into the usage model. In total, 89% of the participants answered the open questions ("What are your wishes concerning infrastructure within the context of electric mobility?") in this section. The question about the necessity of a charging infrastructure yielded a mean value of 1.59 with an available continuum from 1 ("very relevant") to 6 ("not relevant").

Insurance is an important part of the usage model as it is mandatory for motor vehicle owners. The given answers in this section were often rather emotionally phrased which underlines the significance of this subject for electric vehicle owners.

The survey had two questions about *workshops* in the context of maintenance – the topic was also indirectly addressed in the questions about accidents and repairs. As 45% of the users already had to get their car repaired, "workshop" is also a relevant subject matter. On average, the participants answered the question about their satisfaction with the range of existing workshops only with 3.27, in a continuum from 1 ("very satisfied") to 6 ("very dissatisfied"). This shows that the subject is a critical issue with visible deficiencies. It is thus correctly integrated into the usage model.

Production was not specifically addressed in this survey. But the participants frequently mentioned the manufacturer in the open questions: "What would you improve about the possibilities for information?" and "Were you satisfied with the provided information? Why?" The users consider the manufacturer as an obvious contact person for purchase and further development, thus confirming the manufacturer as an actor within the usage model.

Retailers, however, are explicitly mentioned in the survey. The question "What source of information did you use before the purchase?", for example, presented the retailer as an answer option. Nearly half of the participants stated that they had turned to car dealers for information. Another question asked about the number of car dealers the participants had contacted before the purchase. The result was a mean value of 2 – the widespread Internet usage qualifies this as a rather high value.

Other open questions also produced various statements on the topic of retailers. The question about ways to improve the possibilities of information was one of them. 39 participants considered this to be a responsibility of the retailers. 24 participants named good or bad retailers as a reason for their rating of the provided information. This makes the retailer a relevant actor within our usage model.

A very open question addressed the *workplace* ("What kind of interconnectivity do you desire in the context of electric mobility and work?"). This question was surprisingly well received – 78% of the participants answered it, and the statements were often long and extensive. The importance of the workplace was emphasised multiple times in the open question "Concluding Comments". As the participants also wished for a more accurate grasp of the subject "workplace" and further elaboration on the subject, we incorporated the "workplace" into the usage model.

Internet platforms were given as an answer option to the question "What would be your preferred way to resell your electric vehicle?", and chosen by 62% of the participants. 80% of the participants stated that they had used the Internet as a source of information before the purchase. The desire for an app, which was expressed in the answers to the open question about infrastructure, also verifies that the Internet is relevant in the context of electric mobility. Thus, the Internet is a very global factor in the usage model.

The survey contained no specific questions about *carsharing*. However, various participants mentioned the subject in the open questions "What kind of interconnectivity do you desire in the context of electric mobility and 'living'?" and "What kind of interconnectivity do you desire in the context of electric mobility and work?" As some of the answers were also quite extensive, we included the topic in the usage model.

The answers in the context of *resale* showed that only 8 participants (3,7%) offered no information on how they would re-sell their car. This minimal number shows, that most users consider it a relevant subject and give it some thought, i.e. that resale is a part of the usage model. However, resale is not included as a separate actor, because 99% of the users would use the Internet or a car dealer (retailer) for the resale of their car.

A hitherto unexpected part of the user system is legislation and thus the government. The government is not incorporated as a separate actor, because all wishes and shortcomings fall into areas that are already included. In addition to financial investments into a reasonable infrastructure, the users want statutory regulations. Some electric motorists also support special rights and privileges for electric vehicles, such as reducing the vehicle tax, free parking, using the bus lane and/or other privileges.

Thus, all areas can be derived from the high percentage of answers to direct questions or the frequency of responses to open questions. Furthermore, the areas we detected can be logically linked to electric vehicles.

5 Open Innovation Survey

In the further course of the usage-perspective approach, we conveyed an open innovation survey to validate the usage-phases we had acquired so far. The survey was broadly structured into four sections. The survey started with the demographic data and contained multiple-choice questions as well as questions for free answering. The following three sections had a similar structure. They presented a broad title in combination with various aspects/items and a large text field for free answers, which allowed the participants to write down whatever came to their minds about the combinations. Additionally, each main section title provided a separate field for ideas and thoughts. The three main titles were "Interconnected Mobility (Smart Mobility)", "Car Trade" and "Public Transport". The survey concluded with four open questions about the connection of public transport and cycling.

5.1 Participants of the OI Survey

The online survey was conveyed over a period of two months. The 32 participants (31% female) between the age of 21 and 62–34 years on average – used the provided text fields in all sections.

5.2 Evaluation of the OI Survey

The open questions were carefully analysed by using the analysis principles of the problem-centred interview and Grounded Theory [7, 8] as a guideline. The objective of the problem-centred interview and also of Grounded Theory is an unbiased capturing of subjective opinions. The interview guideline or online questionnaire is used as an orientation guideline for the participants. Thus, the data ascertainment is not influenced by the formation of ex ante categorisations, which is crucial for an objective theory approach. The analysis consists of open coding, axial coding and selective coding. The explorative procedures meets the objective of an unbiased opinion survey, and makes it possible to include new, innovative ideas.

5.3 Results of the OI Survey

The survey results showed a new requirement perspective on mobility, which was present across all topics: users want an access mobility that is as standardised as possible. The users want one 'logic' and an interaction scheme within the various "many-to-many" mobility relationships. It is not important to them whether their demands are met by the online presence of public transport companies, the city or other mobility provider. They only want it to happen in a standardised, fast, holistic, mobile and multimodal way. Thus, users only finally demand an implementation of the opportunities that digitization and globalisation have made possible. But next to an adaption of the digital world to their needs, the users want an adaption of existing, real infrastructures to the digital world, e.g. with regard to usage and billing combinations of public and individual transport. This also creates new business models and processes, such as "flat rates for mobility". Because in the end, it only matters that we arrive at our destination in a reasonable amount of time. Travel time, availability, comfort and constancy define the travel costs. Nevertheless, users want central, topic-related platforms (PaaS and SaaS) that bundle, summarise and structure the solutions for their needs in a usable fashion.

6 Expert Interviews with Aspiring Engineers

The above results of the open innovation survey were presented to a focus group of engineer students. The results were discussed with an average of five participants, one moderator, one secretary and another observer in three consecutive sessions. All participants are at least close to receiving their undergraduate degree in either information technology or electrical engineering – more than half of them, however, are already working on their graduate degree. It was important that all participants had already completed the majority of their degree in order to be as close as possible to the end of their education and about to enter the workplace. Additionally, only students were selected that had a strong interest in R&D. All participants of this focus group want to develop their first products in the coming years, or at least aid the development of products. The reason for choosing this form of focus group was to implement a resource saving "quick and easy" [9] discussion approach with rapid and versatile results. Like the open innovation survey, the focus group survey is conducted with a guideline and cooperation model [10]. The results of the open innovation survey were used as a guideline for the focus group and discussed one after the other. The evaluation of the results was limited to the group [11] and describes the various opinions and explanations with regard to the topics of the guideline. The evaluation solely considers whether the participants produced any results and to what degree those results fit the previously made statements. Group specific processes and dynamic effects within the focus group are not part of the research and are thus not considered.

6.1 Results of the Engineer Interviews

To observe the evaluation of the results by the students, the moderator started each session with a short introduction and then offered one topic for discussion. Very quickly it became clear that the task was very abstract, and that the students were only able to deliver a few or no evaluation criteria at all with the exception of personal opinions. Thus, the moderator already intervened after a short period of time in the first session and started to construct a real situation in companies. All participants were asked to assume the role of an engineer inside this company, primarily in the development department, and to present new ideas to the board or management, and convince them that the new idea was of use, feasible, relevant and promising. Additionally, the students had to name the number and expertise of the required employees. For every further result, the student switched characters and were supported, supplemented and occasionally rectified by the other group members. The executing staff perceived a great deal of insecurity among the students. The students were unable to deal with the abstract task and almost exclusively relied on their personal expertise in the more practical examples. For the selection of the required staff, the participants primarily chose members of their own departments regardless of the given task. For example, the electrical engineer thought that the implementation fell into his competence whereas the computer scientist was of the opinion that it fell into his. In addition to their own expertise, the students named marketing as being necessary to implement ideas. After two of three focus group sessions, the researchers realised that the aspiring engineers would not proceed in a different way. The students always regard the development process from a self-centred perspective and are unable to differentiate even within a group consisting of people of different fields of expertise. The investigators executed the third session without any customization or alterations in the evaluation of the survey. After a renewed confirmation, they assumed that further sessions would not show a significant shift.

A further result of the focus group is that the focus group elaborated very similar approaches for every task. It must also be acknowledged that the participants were only able to perform a kind of "cognitive walkthrough" after the intervention of the moderator. They were only able to evaluate the feasibility of the usage-specific survey results – and in some rare cases to alter them accordingly – when they were presented with the practical example of a potential future work situation.

The statements about relevance, feasibility and chance of success were made almost exclusively from a technical point of view. Chances of success were repeatedly seen as falling into the domain of marketing and not of engineering. The investigators deliberately renounced to lead the conversations into a certain direction in order to avoid any learning effects in rapidly successive tasks.

7 Results

In the context of research of activity processes in early development phases it was observed, particularly via the conducted interviews with aspiring engineers, that relevant beneficial potentials could not be assessed or were not identified at all due to the mono-disciplinary education. Every participant of the entire research lacked different information, experiences and methods in order to make reliable assertions about the newly identified potential solutions. Solely technical viable requests could be evaluated based on their technical feasibility. In order to remain accessible in the digitization of increasingly complex systems, or even help designing them, the education of engineers needs to adapt in the medium term and pursue interdisciplinary approaches. This is particularly the responsibility of colleges and universities that are challenged with providing new interdisciplinary and constructive learning approaches [12].

If it is possible to teach engineers the procedures of modern business models and processes, and how to synthesize useful beneficial potentials useful and how to compare those two [13], engineers will be able to make their own rudimentary decisions and/or consider or scrap requirements in early development stages.

At the moment, the expertise of evaluating ideas effectively and efficiently already at the very beginning is not taught in this form. That is why there is currently no recommendation or methodology to objectively, reliably and repeatedly derive and evaluate requirements for the development of products, systems or services so that those withstand even when considering economic components.

In contrast to invention [14], which solely consists of the idea, the solution and the invention (the creation of something new), this ascertained but not yet defined expertise and the newly identified type of engineer proceed according to a specific procedure in order to evaluate an invention objectively, as fast as possible and cost-efficiently. On the other hand, this new type of engineer will not be responsible for the marketable implementation and commercial exploitation of the positively assessed and prototypically implemented invention. This type of engineer is thus not to be equated to the entrepreneur [15], who is responsible for the external economical exploitation, nor to the intrapreneur [16], who is responsible for the further internal development and optimization of the invention. Intrapreneurship is defined as "a spirit of entrepreneurship within the existing organization" [17]. It is very uncommon that inventors are also entrepreneurs, if so it is most likely by chance [14]. The type of engineer that fits between inventor and entrepreneur is, as of now, defined as the "innovateur".

The accomplishment to evaluate and further develop ideas regarding miscellaneous technological and economic factors cannot emerge from within the user. Like specialists, users are far too mono-disciplinary – interdisciplinary thinking can thus not be assumed. Furthermore, the average user does not posses the required technical know-how, the state-of-the-art technology or has information about competition or possible company internal problems regarding development.

Figure 4 shows in isolated steps how the usage-perspective approach works. At first, all issues regarding the usage phases of the product are identified and the corresponding personas, scenarios and use cases are derived. They are used for methods, which are further processed by specialists. After customizations, functional and usage beneficial requirements are described and are partially used for specification via mockups etcetera. The requirements are developed ready for production in subsequent development processes, and evaluated according to usage-based needs before they are specified as mandatory.

Thus, the developer ("innovateur") should perform the process of evaluation. The "innovateur" needs to grasp the entire structure of the idea and to understand the development process from multiple perspectives. In addition to the technological perspective, the "innovateur" should also be able to consider usage-centred and economical aspects as well. At the beginning, the usage should always be sketched in order to regard the real user requirements objectively, and not limit themselves through different perspectives right away.

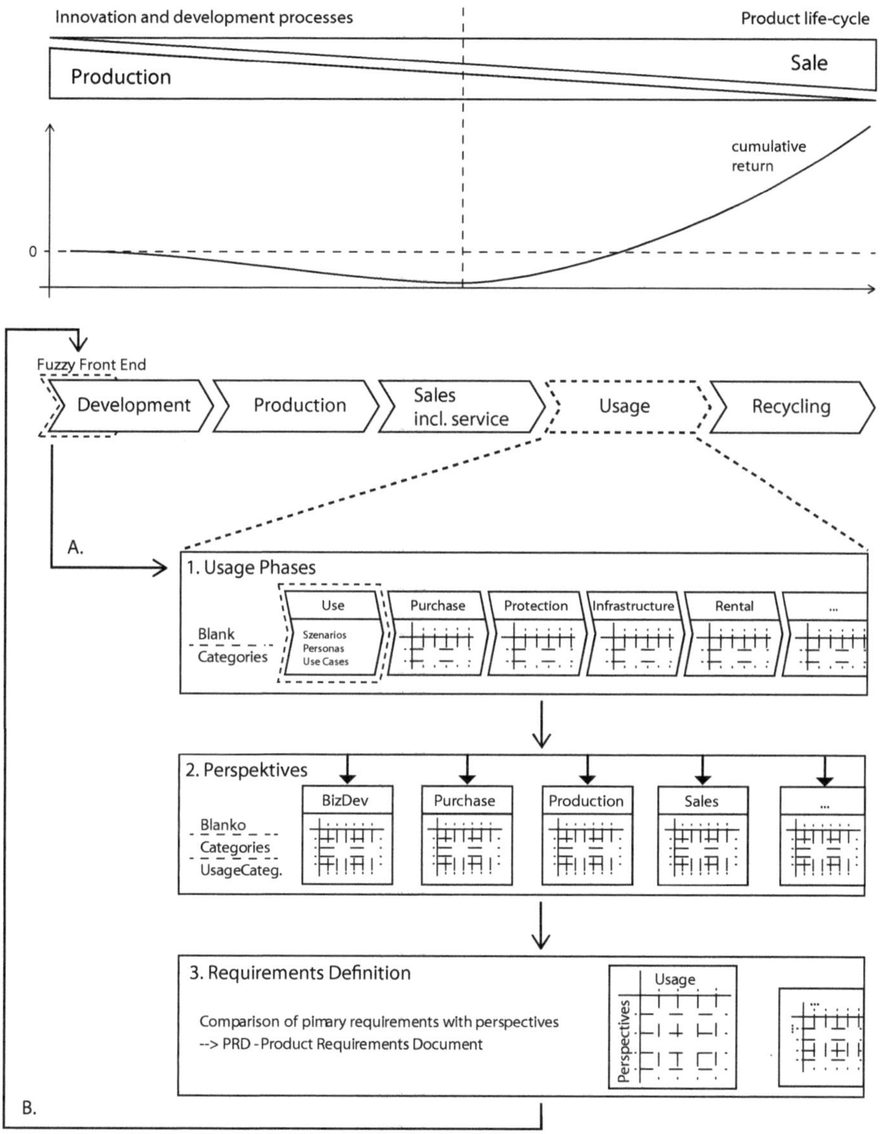

Fig. 4. Usage-perspective development approach

8 Conclusion

The previously examined usage categories allowed us to identify and compare relevant perspectives of the product-life cycle in order to increase the usability [18] and revenue of a solution and to control the development processes. The evaluation of an innovation with broad core function tests by specialists (perspectives) and an associated feasibility study by an "innovateur" promotes a measurable increase in implementation quality.

Offering a process model based on usage phases and perspective shifts to the actors involved in the development process, can increase the quality of the planned solution. The interface between requirements definition and subsequent development is of particular importance. It presents a major challenge in standardisation. Further research should elaborate reliable implementation procedures that facilitate the information transfer of the requirements and the corresponding solutions and also reduce the adjustments. The usage-perspective development approach proved effective in the practical development [19, 20] of a smart home and mobility platform (hardware, mobile and web portal). It will be used in the early development stages (fuzzy front end) of further adjacent issues, and thus examined more deeply. The application of the method is also considered helpful for industrial activities, thus allowing it to investigate corresponding questions and issues in further works. The specification of the approach is thus particularly suited for a practical application in research and development. It should be able to achieve a noticeably reduced workload and increase effectiveness and efficiency. The prospect is the field evaluation of the new construct in a usage-based and role-based perspective structure, and its incorporation into a modelling and design software for early development phases.

References

1. Schimanski, S.: Usage phases in the development of product systems exemplified by a route recommendation scheme for cyclists. In: Human Computer Interaction and Transportation, HCI International. Springer, Toronto (2016)
2. Endsley, M.R.: The role of situation awareness in naturalistic decision making. In: Zsambok, C.E., Klein, G. (eds.) Naturalistic Decision Making, pp. 269–283. LEA, Mahwah (1997)
3. Schimanski, S.: Zur Organisierbarkeit von Ideenpotentialen. Untersuchung ambivalenter Managementansätze der frühen Innovationsphasen. Unpublished thesis, Wuppertal (2004)
4. Schimanski, S.: Production ergonomics – analysis methodology for human-system-integration in teamwork. In: Experimental Industrial Psychology IV: Human-Computer-Interaction, Proceedings of 16th EAWOP Congress – European Association of Work and Organizational Psychology, Münster (2013)
5. Verworn, B., Herstatt, C.: Prozessgestaltung der frühen Phasen. In: Management der frühen Innovationsphasen, pp. 195–214. Gabler Verlag (2003)
6. Rasmussen, J.: Information Processing and Human-Machine Interaction. North-Holland, New York (1986)
7. Strauss, A., Corbin, J.: Grounded Theory: Grundlagen Qualitativer Sozialforschung. Beltz, Psychologie Verlags Union, Weinheim (1996)
8. Witzel, A.: The problem-centered interview. Forum: Qual. Soc. Res. **1**(1) (2000). http://www.qualitative-research.net/index.php/fqs/article/view/1132/2519
9. Morgan, D.L.: Focus Groups as Qualitative Research. Sage Publications, Thousand Oaks (1997)
10. Alexander, T., Schimanski, S.: Modeling cooperation and communication processes within infantry teams. In: SAE International, Paper #08DHM-0033, Proceedings of the 2008 SAE Digital Human Modeling Conference, Detroit (2008)
11. Schulz, M.: Quick and easy!? Fokusgruppen in der angewandten Sozialwissenschaft. In: Fokusgruppen in der empirischen Sozialwissenschaft, pp. 9–22. VS Verlag (2012)

12. Reich, K.: Konstruktivistische Didaktik: Das Lehr-und Studienbuch mit Online-Methodenpool. Beltz (2012)
13. Giddens, A.: The Constitution of Society, University of California Press (1984)
14. Schumpeter, J.A.: Theorie der wirtschaftlichen Entwicklung, Berlin (1911)
15. Bull, I., Willard, G.E.: Towards a theory of entrepreneurship. J. Bus. Ventur. **8**(3), 183–195 (1993)
16. Antoncic, B., Hisrich, R.D.: Clarifying the intrapreneurship concept. J. Small Bus. Enterp. Dev. **10**(1), 7–24 (2003)
17. Hisrich, R.D., Peters, M.P.: Entrepreneurship: Starting, Developing, and Managing a New Enterprise, 4th edn. Irwin, Chicago (1998)
18. Herczeg, M.: Software-Ergonomie: Theorien, Modelle und Kriterien für gebrauchstaugliche interaktive Computersysteme, vol. 3. Auflage, Oldenbourg Wissenschaftsverlag (2009)
19. Schimanski, S., Schroeder, T.-F., et al.: Contactless sensor technology for measuring soil moisture. In: Smart Home-Solutions – Network-Interface-Technology, ICCE - International Conference on Consumer Electronics, Las Vegas. IEEE (2014)
20. Schimanski, S., Runge, S.: Optimization approach and using Bezier-splines in navigation algorithms. In: Automotive, ICCE - International Conference on Consumer Electronics, Las Vegas. IEEE (2015)

Human Factors Approach to Study Border Control Automation Impacts and Needs: Methodology and Preliminary Results of Field Studies

Elise Le Guellec[1(✉)], Christine Megard[2], Grigore M. Havârneanu[3], Sonia Huguenin[2], Virginie Papillault[3], Juliette Richoux[1], and Sirra Toivonen[4]

[1] Thales Communications & Security S.A.S, 4 Avenue des Louvresses, 92230 Gennevilliers, France
{Elise.Leguellec,Juliette.Richoux}@thalesgroup.com

[2] Institut CARNOT CEA, LIST, DIASI/Laboratoire d'Interfaces Sensorielles et Ambiantes, CEA Saclay Nano-INNOV, Point courrier 173, 91191 Gif sur Yvette Cedex, France
{Christine.Megard,Sonia.Huguenin}@cea.fr

[3] International Union of Railways, 16 rue Jean Rey, 75015 Paris, France
{Havarneanu,Papillault}@uic.org

[4] VTT Technical Research Centre of Finland, Tekniikankatu 1, 33720 Tampere, Finland
Sirra.Toivonen@vtt.fi

Abstract. Passenger flows are continuously increasing in Europe and the number of border guards does not increase as quickly as it needs. The use of automatic systems such as e-gates and kiosks is envisaged to enhance security and to facilitate the border crossing. Border control activity should be thoroughly studied in order to understand in which ways it would be impacted by the introduction of more technological systems. The purpose of this study is to analyze the current border guards' activities from a human factor point of view and to provide recommendations and requirements regarding the introduction of the future regulation and the use of automatic systems. The paper introduces the methodology used to investigate human factors at four types of borders based on a systemic human factors approach, organizational factors, technical tools and environmental aspects.

Keywords: Automated border check · Border guards · Smart Borders · Europe · Man machine allocation · Cognitive models · Recommendations

1 Introduction

As traveler flows are continuously increasing in Europe, the current security checks at the European Union (EU) external borders can be very long and challenging. The new forecast published by Airbus [1] anticipates a air traffic growth of 4.5% annually and the number of Schengen border crossings in 2025 will possibly reach 887 million with one

T. Ahram and W. Karwowski (eds.), *Advances in Human Factors, Software, and Systems Engineering*, Advances in Intelligent Systems and Computing 598,
DOI 10.1007/978-3-319-60011-6_2

third concerning Third Country Nationals (TCNs: any person who is not a EU citizens). At the same time, the increase of the number of border guards (BGs) will certainly not be able to cover the needs from the field. The EU wants to enhance security at the border toward TCN and to deploy means to track over stayers. The Smart Borders Initiative from EU plans to introduce an Entry/Exit system (EES) that would record entrance/exit dates and places in the Schengen area for TCN and data regarding any refusal of entry. The use of automated systems such as e-gates and kiosks is also envisaged to enhance traveler's flows at the border and to "facilitate" the border crossing of TCNs.

Border control is described in the Schengen code [2] by four main subtasks: check if the identity of the traveler matches with his/her identity document, checks the authenticity of the identity document, check the validity of the identity document and check if the traveler can be a threat. The study is focused on the activity at the first line which is the first check performed by border guards. In case of a doubt, the traveler is conducted to a second line. This first line activity should be especially considered in order to understand in which way it could be impacted by the introduction of more technological systems and automation. To our knowledge, the activity of border control has rarely been the subject of human factors publications. Despite the fact that the impact of automation on human factors has been largely addressed in the past three decades [3–5], most of the studies were related to disruptive technologies like cockpit automation. Very few were focusing on border control, except some on the impacts of e-gates on travelers [6].

The purpose of this study is to analyze the current border guards' activities from a human factor point of view and to provide recommendations and requirements on the introduction of the future regulation with EES and the use of automatic systems at the border.

This paper introduces the human factor methodology used in the EU research project BODEGA. The methodology is based on observations and interviews of different actors involved in the border control. The approach is based on a systemic view of border control activities, organizational factors and technical tools. The results of the field studies on these three dimensions will be presented. The analysis is partly based on Parasuraman and Sheridan's taxonomy of mental workload, situation awareness and skills degradation [5]. The paper ends with a discussion regarding the human factor issues that need to be considered for the implementation of the smart border initiative in Europe.

2 Methodology

The methodology used to identify needs and build up recommendations relative to present and future border check is based on a systemic human factor approach considering the entire end-users, from top management and trainers, to border guards and travelers. Other elements of the system like customs, port authorities, airlines, security, body control, sellers, have been considered regarding their relationship with border control. The study focuses on the analysis of the global activity (tasks performed by border guards), organization (team organization, interactions, etc.) and technologies used at border control. The tasks analysis focuses on the tasks performed in real settings. This framework, presented in by Keyser [7], relies on numerous studies indicating that the work is rarely performed as prescribed. In real conditions of work, operators need to adapt the procedures with external constraints (high or very low passenger flow, weather conditions, technologies failure, etc.) and internal factors like motivation, stress, boredom, routine and many other factors that may influence the way to perform their work.

It is very important to catch the reality of the field to be able to provide recommendations and requirements for Smart Border Initiative matching with the current needs. Tasks analysis contains two main steps:

- The first step is the identification of the prescribed tasks and the goals that border guards have to achieve. The tasks define what the border guards have to do and in which order, to be able to achieve their goals. The prescribed tasks can be assessed thanks to official documents (for instance the Schengen code) and interviews with managers.
- The second step is the identification of the real tasks performed on the field and the understanding of the gap reasons between the prescribed tasks versus the real one. Observations and interviews are the main methodologies used to assess an overview of the whole activity.

To achieve the goal to build a real vision of border guards' activity, field studies have been performed across Europe at each type of border crossing points: air, rail, land, port. For a collective analysis of the data, the BODEGA consortium partners defined a common methodology for data collection and analysis. The methodology was more focused on a qualitative collection rather than quantitative to get an in-depth view of the activity of border guards. It included:

- Interviews of border guards, managers, top managers, and trainers.
- Observations of border guards' activity.
- Observations of travelers at the border crossing point (behavior, interactions with BG or automated system …), travelers' interviews.

The semi-structured interview guides included nine main categories which were built according to the goal of BODEGA project:

- Organization and management of the team: description of the organization in the team.
- Activities: description of how border check is performed concretely on the field and description of the task of border check in the legal framework of the Schengen code.
- Performance: factors that affect performance at border check.
- Soft Skills/Non-Technical Skills that contribute to enhance performance at border check.
- Training: description of how training contribute to the efficiency and effectiveness during border checks.
- Evolution of the BG's job and how border guards perceive their role in a prospective view.
- Environment Technology: description of the tools used at border check.
- Travelers view point: perception of border check by travelers.
- Ethics and legal framing.

All participants in the field studies were voluntary. They filled in a consent form accepting the confidentiality of the collected data. It is important to highlight that it has not always been easy to have access to the field due to several issues: time constraints, availabilities of the border guards, number of border guards in the border crossing point, recent terrorist attacks in Europe in particular in France and in Belgium. Several planned field studies were cancelled because of these events and others were post-poned. Below is the overview of the countries where field studies have been conducted: Belgium, Finland, France, Greece, Italy, and Poland (Fig. 1).

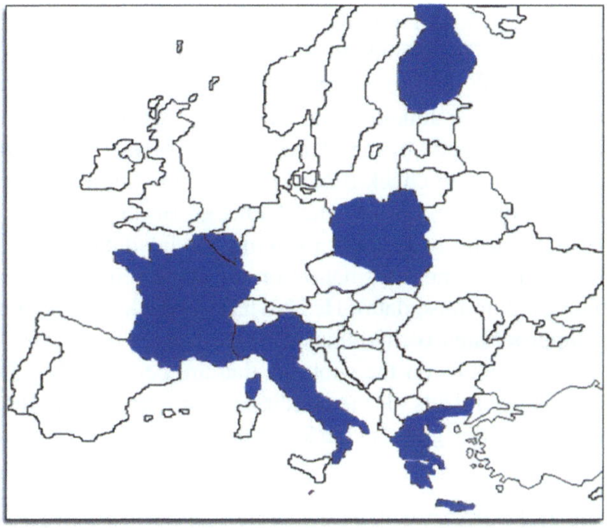

Fig. 1. Overview of the countries where field studies have been performed.

The data collected from the field studies were integrated in an Excel sheet and grouped into the nine main categories (mentioned above) for further content analysis.

The data were coded to keep them confidential. For example, only the type of border (air, road, rail, and maritime) and the profile of the person (gender, age, etc.) were mentioned.

The qualitative data analysis was performed by profile and categories. Generally, the types of borders have been analyzed altogether but specificities were added when needed.

3 First Results

3.1 Tasks Performed at Border Check

This section allocates the four prescribed subtasks, required by the Schengen code, into five main tasks that the border guards shall perform in their activity.

- Risk assessment performed beside border check: the observations at some borders have shown that border guards can contribute to risk analysis. They analyze the lists of travelers for a check in the databases before they arrive at the border. They proceed to an in-depth check of travelers if required. Border guards can also report statistics about stolen documents or falsification of documents.
- Risk assessment regarding the traveler: this main task contains two subtasks. The first one is relative to the appreciation of risk by the border guard before the traveler arrives in front of the BG. This covers broad observations performed before border, like waiting area, or in the queue. The other one relate to the risk assessment during the direct interaction with the traveler.
- Traveler identification: border guards check if the identity of the traveler matches with the information of his/her identity document (photo, nationality, …). Border guards also verify if the data related to the identity of the traveler have not been modified. This main task is linked to the task in line with document verification but is strictly related to the check of the identity of the traveler. In addition to this task, the border guard has also to check if the traveler is not subject to an arrest warrant.
- Document verification: the border guard checks of the validity (date) and the authenticity of the identity document (looking for falsifications or forgeries). The border guards also assess the information on the document (sometimes a document is authentic but based on false data). He also has to check if the identity document is not stolen, misappropriated or invalidated.
- Eligibility check: this relies on the subtasks that are performed by border guards to ensure that the traveler fulfills all the conditions for entry/exit in the Schengen area for TCN travelers. Border guards check the databases to verify that the traveler is not a threat.

All these main tasks performed in the border guards' activity contribute to the decision making regarding the traveler.

The model presents in a time line, the different types of tasks, considering all steps of the process, from pre-risk assessment to decision making. The model shows how the general risk assessment regarding travelers before they arrive in front of the border guard contributes to the risk assessment of the individual traveler who is checked. The pre-assessment allows the border guard to adjust the level of risk if he detects 'something strange' in the behavior of the traveler. In that situation, the border guard will pay more attention during the check (see Fig. 2).

The model shows that the risk assessment with the traveler is performed during the all activity, the entire check and that each mandatory task related to the check (checks of the identity; check of the authenticity of the identity document and check the eligibility conditions) contributes to adjust the evaluation of the risk associated to the traveler. The tasks performed at border check contribute to decision making. The information gathered through the consultation of the databases, the check of the visa (with the Visa Information System (VIS)) or the calculation of the traveler in the Schengen area (for TCN only), also contributes to border guard's decision about one traveler.

The model supports the idea that the duration during which the border guard interacts with the traveler is of primary importance in the quality of the control.

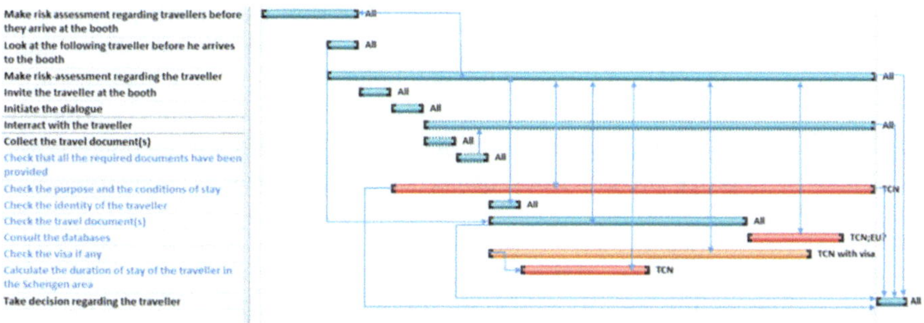

Fig. 2. The model shows how the activities performed during border check in first line contribute to re-assess the risk of each traveler and to decision making.

3.2 Organizational Aspects

The field studies also highlighted different levels of organization that appear to be important in the management and functioning of the border control points (BCPs). One of the most important finding is that the organization of BCPs is different across various countries, border types and BCP sizes, but there are also common organizational aspects such as: the manager's role, teamwork, collaboration, communication and exchange of information.

The local management of the team and the role of the manager are very important to organize the team, to encourage and motivate the BGs, to manage the information to the BGs, to give enough feedback about the BGs performance and give them useful information for their job. The team management needs to take into account versatility, flexibility and complementarity of the staff. The versatility of the BGs allows managers to assign them adequate roles in the work place. While some BGs would like to have specialized roles, others prefer to be more versatile. It seems that BGs are motivated when the organization allows them to perform several activities: first line checks, second line checks, writing reports and minutes – in other words being responsible for the whole process from the beginning until the end.

In the organizational process of a BCP it is also crucial to consider the various interactions existing between the BGs, especially the way they cooperate, the way they communicate together and finally the way they exchange information or share their experience. Teamwork, collaboration and communication are part of the BGs work and enable them to: be complementary when they work in the booth, help each other when they face an unusual or difficult situation and be aware of all important information they need to perform their job. Teamwork is also important because BGs are more complementary in pairs and can help each other, allowing informal sharing of information about encountered situations. For example, it gives them the possibility to ask for advice, opinions (e.g. about an incident, a document) and to help their teammate in case of doubts or in case of technical problems.

Team cooperation and exchange of information with other services are equally important. Communication exchanges between BGs helps them identify errors and learn from the experience of others. Sometimes informal meetings and debates are organized to share their experience about specific cases. BGs work generally with many other services, depending on the type of BCP. In most cases these are the customs officers, or the BGs of the neighboring country. We noticed a huge disparity in the cooperation between BGs and the other services; some work closely, others not. Finally, communication and dissemination of information between the different BCPs are essential to improve the efficiency of border check especially concerning neighboring countries.

3.3 Technological Aspects

In the last years, border guards have seen several introductions of new technological tools in their job (inside booths); inter alia document reader and fingerprint recognition who have contribute to improve the control efficiency. Border guards appreciate these tools because they make them feel more confident about their decisions (let the traveler cross the border or not).

Smart Borders and the following establishment of the EU Entry-Exit system will be the next big (r)evolution (decision of the proposal is expected spring 2017 [8]). The proposed changes include additional biometrics verification, manual stamping withdrawal and the possibility for TCN citizens to use automated border checks systems (so far reserved for European citizens only). In summary, the proposal is expected to introduce more self-service kiosks and e-gates at the different border types.

Kiosks are not deployed yet in Europe but e-gates have been used in the largest European airports for years. Most of the fields observed did not have automated systems deployed and border guards had no information about the future evolutions expected. For this reason, they think that it is not possible to replace their knowledge, their skill and their expertise by machine. A main question is related to risk assessment activity that is a main activity of the border control process. This activity seems to be the most difficult to automate.

Technology failures could be an obstacle for the adoption of technological tools. If BGs consider that a tool adds more work, for example it makes the control time longer, they may not use the tool.

For border guards, having more technologies could be useful but they still want to keep control. For instance, they would like to have more technologies embedded inside the manual booth or kiosks before the booth control. In this way, tools help them to take decisions (confirm or not their first doubts about a traveler). In other words, border guards do not think that it is possible to be replaced by machines. For them, the final decision should stay human, which is also in line with the Smart Borders legislation proposal.

The main recommendations are that the technology deployed on the field should be mature and reliable enough to be accepted and used by border guards. Sometimes, the organization are not enough aware about the importance to provide a maintenance services in addition to the deployment of new tools.

Regarding automated systems, border guards' activity is affected. From an active position (control the travelers), they will be more into a passive one (supervise the automated control and react in case of problems). Though, it is well agreed that automated machines always require human oversight. In future, border guards will probably operate more like alert spotters of anomalies rather than frontline passport-checkers [9].

The deployment of Smart Border should take into account: training, change management and to define a good balance to keep the border guards in the loop of the border control.

4 Discussion

During the last years, the border environment has evolved significantly with a more and more consequent passenger flow and the amplification of new risks such as terrorism. One border guard may check thousands of travelers weekly and he has a very short period of time to decide if the traveler is legitimate or not, and to assess if the traveler's documents are genuine. Border guards also should take into account urgent inputs regarding risk analysis for illegal immigration and the security of the member states.

From a distant viewpoint, border guard's job may seem boring due to the repetitive tasks to perform. However, the study showed that border guards' activity is more complex than just stamping passports. They find motivation in contributing actively to the security at the borders and in interacting with a diversity of travelers. Border checks are performed in close interaction with people from all over the world, speaking different languages, and with people from different religions, cultural habits and behaviors. These interactions are actual clues that help border guards to take decisions about travelers. It is an important part of the risk assessment activity.

One important thing learnt on the field was the importance of experience in border guards' job. They have basic training to learn about border control, but most of their skills are learned on the field thanks to the practice and the interactions within the team. Their skills are mainly composed by soft skills, which were hard for them to explain.

The results of the study highlighted the diversity of the processes between the different types of borders and the necessity for the border guard to be flexible enough to deal with the changing situations in their daily activities. The introduction of new technologies should be studied by taking into account the type of borders but also the infrastructure constraints, which appeared to be important issues. There cannot be enough space to install e-gates, therefore self-service kiosks seem to be more adapted in these cases. The traveler population characteristics should be also considered because they are not the same between:

- Road and rail borders with lot of daily travelers and few nationalities.
- Air borders with lot of nationalities.
- Maritime borders with many elderly travelers who need assistance and are not so comfortable about using technology by themselves.

With the introduction of automated systems, travelers diversity should be considered and the deployed systems should be adapted for most of them.

The study showed that the change management and the training of border guards are not always following the introduction of new technologies, even though they are important to ensure that the technologies are understood and used in the proper way.

The level of the border guards' trust in automated systems also needs to be addressed. The machine has to be a tool to help them. Border guards should keep their role on the risk assessment that they are currently performing during the manual control.

The tasks allocations between man and machine in EES implementation should be carefully studied before the deployment of e-gates and kiosks, because they can affect border guards' activities, border security, technology implementation and finally also the organization's activities.

To go further in this analyze regarding new technologies, it is necessary to go on additional fields which use automated systems because the main observed fields during the first part of this study do not have yet any automated technologies deployed.

Acknowledgements. This work was conducted under the BODEGA project (Proactive Enhancement of Human Performance in Border Control; http://bodega-project.eu). This project has received funding from the European Union's Horizon 2020 research and innovation program under grand agreement No. 653676.

References

1. Airbus (2016) http://www.airbus.com/company/market/global-market-forecast-2016-2035/
2. Schengen Code: Regulation (EU) 2016/399 of the European Parliament and of the Council of 9 March 2016 on a Union Code on the rules governing the movement of persons across borders (Schengen Borders Code). ELI: http://data.europa.eu/eli/reg/2016/399/oj
3. Bainbridge, L.: Brief paper: Ironies of automation. Automatica **19**(6), 775–779 (1983)
4. Parasuraman, R., Riley, V.: Humans and automation: use, misuse, disuse, abuse. Hum. Factors **39**(2), 230–253 (1997)
5. Parasuraman, R., Sheridan, T.B.: A model for types and levels of human interaction with automation. IEEE Trans. Syst. Man Cybern. Part A Syst. Hum. **30**(3), 286–297 (2000)
6. Oostveen, A.-M.: Non-use of automated border control systems: identifying reasons and solutions. In: HCI 2014 - Sand, Sea and Sky - Holiday HCI Proceedings of the 28th International BCS Human Computer Interaction Conference (HCI 2014), Southport, UK, 9–12 September 2014
7. Keyser, D.: Work analysis in French language ergonomics. Ergonomics **34**, 653–669 (1991)
8. COM (2016)194: Proposal for establishing an Entry/Exit System (EES) to register entry and exit data and refusal of entry data of third country nationals crossing the external borders of the Member States of the European Union and determining the conditions for access to the EES for law enforcement purposes and amending Regulation (EC) No. 767/2008 and Regulation (EU) No. 1077/2011. http://eur-lex.europa.eu/legal-content/EN/TXT/?uri=CELEX%3A52016PC0194
9. Fergusson, J.: Twelve seconds to decide. Frontex (2014). http://frontex.europa.eu/assets/Publications/General/12_seconds_to_decide.epub, ISBN 978-92-95033-92-4

Strategies to Improve At-risk Students' Learning and Achievement: Undergraduate Software Engineering Course Case Study

Suad AlRamouni(✉) and Hisham Alkhawar

Prince Sultan University, Riyadh, Saudi Arabia
{salramouni,halkhawar}@psu.edu.sa

Abstract. In this research paper, we have investigated how to improve at-risk students' learning and achievement in Software Engineering-design and development course, which is an undergraduate course that combines both knowledge of software development and practical skills for analysis and design of software projects. The study covers three main issues: how to early identify at-risk students in the class; how to engage at-risk students positively in the class activities and how to pay more attention to at-risk students in the class. Several strategies to improve at-risk students' learning and achievement were recommended: adjust teaching style to combine meaningful, engaged learning teaching style and adapt to Constructivist teaching and learning models; design challenging class activities that are meaningful to the students that correspond to real-world tasks and require collaboration with peers in the classroom as well as with others outside the classroom; design assessment of learning method in terms of presenting a real-world project and collaborating with small size team and make use of technology to design interactive instructional environment.

Keywords: Software engineering · At-risk student · Assessment · Constructivist

1 Introduction

The purpose of this study is to investigate how to improve at-risk students' learning and achievement in Software Engineering-design and development course - which is an undergraduate core course in the computer science program at Prince Sultan University. To define at-risk student, Maxwell [1] states that this group of students' skills, knowledge, motivation, and/or academic ability are significantly below those of the 'typical' student in the college or curriculum in which they are enrolled. In addition, Ender and Wilkie [2] state that these students are likely to display any number of other characteristics such as low academic self-concept, unrealistic grade, extrinsic motivation, low self-efficacy, inadequate study skills for college success, a belief that learning is memorizing, and a history of passive learning.

The general goal of the software engineering education community is to provide a pedagogically sound framework for educators [3]. One of the main issues which are under discussion in the community is alternative ways and strategies of teaching [4]. Among other proposals, the importance of practice-based software engineering

T. Ahram and W. Karwowski (eds.), *Advances in Human Factors, Software, and Systems Engineering*, Advances in Intelligent Systems and Computing 598,
DOI 10.1007/978-3-319-60011-6_3

education is well accepted in the community and some good examples are reported in the literature [5]. The importance of project based software engineering as a teaching method is well recognized.

2 Background

Prince Sultan University (PSU) located in Riyadh, Saudi Arabia is the first private university funded in Saudi Arabia by Riyadh Philanthropic Society for Sciences. Initially, it was founded as Prince Sultan Private College in 1999 and received university status in 2003. PSU offers courses and programs leading to officially recognized higher education degrees such as bachelor degrees, master degrees in several areas of study. Software engineering program is one main program offered by the college of computer and information sciences.

The Software Engineering-design and development course was monitored for four consecutive academic terms by same instructor. Table 1 shows the details of each class. Class composition refers to the distribution of students per class according to their learning and achievement level. Three levels were identified: Above expectation; meeting expectation and at-risk students.

- **Above expectation students** are those who acquire a mastery of the skills and complete knowledge prerequisites for successful learning, get high test scores, low absence counts, with very active participation in the activities in classroom.
- **Meeting expectation students** are those who acquire acceptable skills and some knowledge prerequisite for successful learning, get good test scores, low absences counts, with acceptable level of participation in the class activities.
- **At risk-students** are those students who have a higher probability of failing academically or dropping out of the course, get low test scores, high absents count, and rare class participation.

Table 1. Class details

Academic year	Term	# of enrolled students	Class compositions
2013–2014	Term1	12	Meeting expectation = 8 Above expectation = 4 At-risk = 0
2013–2014	Term2	8	Meeting expectation = 6 Above expectation = 2 At-risk = 2
2014–2015	Term1	12	Above expectation = 2 Meeting expectation = 8 At-risk = 2
2014–2015	Term2	16	Above expectation = 3 Meeting expectation = 11 At-risk = 2

During these four terms, we have always considered improving our teaching practices to help and support the at-risk students to improve their learning and achievement. The course under consideration combines both knowledge of software development and practical skills for analysis and design of software projects. To make sure that every student in the class will acquire required knowledge and master the essential practical skills, the following teaching practices were planned and implemented:

- Course objectives were defined.
- Course learning outcomes were defined
- For each defined learning outcome, teaching strategies and assessment methods were defined.
- Topics were selected and mapped to the defined learning outcomes

Teaching strategies includes lectures, discussion groups; in-class tasks and case studies. In addition, Assessment methods include in-class exams; Take-home assignments and teamwork project.

Small groups of 2–3 students were formed at the beginning of the classes to work on the project [6]. Students were given the freedom to form their own teams, select their partners and propose a software system to be developed. The projects requirements span the whole term with phased submission of the completed tasks.

Dynamic small groups were formed every class where different students are assigned to a group. Same task was assigned to all groups and require all members of a group to work collaboratively on the task and a summative result to be produced [8]. The group produced result was evaluated and all members received the same score. Group membership changed every class. There was always an intra-groups competition.

Early in the classes, at-risk student were identified based on their low test scores, rare class participation and playing a watching role only in the groups whenever assigned a task. Some practices have been developed to help improve the at-risk students' learning and achievements including - calling at-risk students for a face-to-face meeting where each student was met individually. In the meeting, we discussed with them their poor performance and try to investigate the reasons and offered a special assistance in terms of extra classes and provide extra supporting materials to help them enhance their learning. We chose some references for self-reading to encourage the students to rely on themselves to improve their learning. A follow-up plan was setup with students to check their progress. We also encouraged them to come back early to me whenever they face any difficulty in their understanding and learning.

Also, we adjusted the team formation for in-class assignments; we assigned the at-risk students to groups having above-expectation students and always kept an eye on those students and encouraged them to participate in the class.

For the team project, members have been assigned different roles and requested summary report of the assigned tasks to encourage the at-risk student to accomplish their assigned tasks and work positively by time while improving their learning and achievements.

The problem that we faced and could not reach any satisfactory solution can be expressed in the level of improvement we have achieved regarding improving the at-risk students' learning and achievements. We succeeded in improving their participation in the class activities but failed to help them improve their learning and achievements to reach an acceptable level.

3 Plan for the Intervention

To improve our practices regarding improving at risk-students learning and achievements, a plan was developed to do the following:

- Make use of my peer review observation by
  1. Review our current practices regarding helping at-risk students.
- Performing an action research on the issue of improving at-risk students in the educational organization.
  1. Reading and performing a literature review of the problem
  2. Coming up with ideas how to adjust my teaching practices to improve the at-risk students' learning and teaching
  3. Set up a plan on how to implement the ideas
  4. Implement the ideas according to the defined plan
  5. Evaluate the degree of improvement achieved in the at-risk student's learning and achievements.
  6. Conclude the result and iterate till goal achieved.
- Set up a an action plan to adjust our current practices to achieve our goal in helping at-risk students improve their learning and achievements
- Investigate how we can utilize more technology in our classroom to serve our goal.
- Engage in professional development workshops to improve our teaching practices.

4 Methodology

An Action plan for intervention was set up as follows:

- Adjust our teaching style to combine meaningful, engaged learning teaching style and adapt to Constructivist teaching and learning models [7] where students are self-regulated; can define their needs in their own learning goals; can evaluate their own achievements and know how to learn.
- Design challenging class activities that are meaningful to the students that correspond to real-world tasks in the real work place and require collaboration with peers in the classroom as well as with others in the real world outside the classroom (i.e. stakeholders)
- Design assessment of learning method in terms of presenting a real-world project that require student to engage in real-world problem solving and collaborating with small size team (2 or more students)

1. Reconfigure small groups according to their purposes
2. Allow student to design their own evaluation criteria for assessing their project artifacts and final presentation

- Make use of technology [9]

1. to design interactive instructional environment
2. maintain multiple sources for information (online and offline)
3. as an interactive tool for both practicing skills and creating multimedia content to communicate ideas about the material under study

5 Reflections on the Intervention

Our Critical reflections on the interventions follow.

- At-risk students need to be engaged in interesting and challenging learning by ensuring meaningful and engaged learning and constructing rich instructional environment.
- Appropriate assessment methods, meaningful class activities and assignments will help at-risk students to make connection between their classroom learning and real-world problems which result in improving all students - including those at-risk - learning and achievements.
- Making use of technology for providing multiple sources of information, creating interactive class activities and multimedia contents and establishing variety of communication channels to support collaboration among students themselves and among teacher and students has enabled the students to shift their roles from passive learners to active learners.
- Designing an real-world project as an assessment project and requesting students to form their own teams to work on the project tasks, collaborating with outside stakeholders, and designing their own evaluation criteria has significantly improve at-risk students performance and achievements while it increases the required time to accomplish the tasks as the students need to visits the real-world workplace to collect data and communicate with the stakeholders. We overcome this problem by allocating more time to accomplish the project so the students feel unstressed.
- Reconfiguring groups according to their purposes did not show any improvements and negatively affects the collaboration among team members as students showed resistance to the change and request to stick to the teams they formed by themselves.
- Pulling out at-risk students to our office to get extra support did not show any progress or improvements as the students preferred to stick to their teams in accomplishing the tasks and welcoming the support during the class time only.

6 Conclusion

Trying to improve our practices through an action research has opened our eyes on the experiences of other teachers/educators who tried to address the same problem. The findings of the action research helped me to adjust my teaching strategies and practices to help all students - including those of at-risk - to be active learners and improve their learning and performance.

Teaching the same course for more than one term does not mean to re-use same material and methodology. Considering the diversity of learning among students has to been taken into account to plan the classroom activity. Although interactive activities require more time and effort to plan and design, it has proven its roles in improving students learning and achievements.

References

1. Maxwell, M.: Improving Students Skills. H&H Publishing, Florida (1997)
2. Ender, C., Wilki, J.: Advising students with special needs. In: Gordon, V.N. (ed.) Academic Advising: A Comprehensive Handbook, pp. 118–134. Jossey-Bass, San Francisco (2000)
3. Afra, A., Rizwan, Q.: The proposed methods to improve teaching of software engineering. Int. J. Mod. Educ. Comput. Sci. **7**, 13–21 (2016)
4. Soska, A., Schroll-Decker, I., Mottok, J.: Implementation of practical exercises in software engineering education to improve the acquirement of functional and non-functional competences: a field report about project-based learning in software engineering. In: International Conference on Interactive Collaborative Learning (ICL), Dubai, pp. 338–345 (2014)
5. Yadav, S., Xiahou, J.: Integrated project based learning in software engineering education. In: International Conference on Educational and Network Technology, China, pp. 34–36 (2010)
6. Holloway, H.: Grouping students for increased achievements. Educ. Leadersh. **59**, 84–85 (2001)
7. Jonassen, D.H., Rohrer-Murphy, L.: Activity theory as a framework for designing constructivist learning environments. ETR&D **47**, 61–79 (1999)
8. Tomlinson, C.A.: The Differentiated Classroom: Responding to the Needs of All Learners. ASCD, Alexandria (2014)
9. Darling-Hammond, L., Zielezinski, M., Goldman, S.: Using technology to support at-risk students' learning. Stanford Center for Opportunity Policy in Education. https://edpolicy.stanford.du/publications/pubs/1241

Bibliometric Analysis of Human Factors Research: Intellectual Structure and Evolution

Jingyi Wang and Wei Shan(✉)

School of Economics and Management, Beihang University, Beijing, China
amanda1249@126.com, shanwei@buaa.edu.cn

Abstract. Human factors remains a recent field of research that is dynamic and changing with technology and social life. This paper aims to conduct a bibliometric analysis of human factors research. We explore research on Human Factors and Ergonomics through SSCI. We find that highly cited papers is a negligible portion of the total. We map the countries and the most relevant approaches as well as detailing the new theoretical perspectives to human factors theory. The analysis conducted uses multivariate statistical analysis techniques in addition to the co-citation matrix to shed light on these issues. The results provide a global perspective of the field, identifying the works that have had the greatest impact, the intellectual interconnections among published papers, and the main research traditions or themes that explored in human factors. Structural and longitudinal analyses reveal changes in the intellectual structure of the field over time.

Keywords: Human factors · Ergonomics · Intellectual structure · Bibliometric analysis

1 Introduction

The term ergonomics first entered the modern lexicon by Polish scientist Jastrzębowski. The two terms "human factors" and "ergonomics" are essentially synonymous. Human factors and ergonomics is the practice of designing products, systems, or processes to take proper account of the interaction between them and the people who use them. The definition by IEA is commonly accepted. The IEA defines human factors as follows: Ergonomics (or human factors) is the scientific discipline concerned with the understanding of interactions among humans and other elements of a system, and the profession that applies theory, principles, data and methods to design in order to optimize human well-being and overall system performance.

Human factors and ergonomics takes account of the "fit" between the user, equipment and their environment. The subject comprise three main domains of research: physical, cognitive and organizational ergonomics. There are specializations under these broad categories. The field of physical ergonomics include visual ergonomics. The field of cognitive ergonomics include usability, human–computer interaction, and user experience engineering. While some specializations run across the three main fields. Based upon the framework, experts apply an understanding of human

T. Ahram and W. Karwowski (eds.), *Advances in Human Factors, Software, and Systems Engineering*, Advances in Intelligent Systems and Computing 598,
DOI 10.1007/978-3-319-60011-6_4

factors to the design of equipment, systems and working methods in order to improve comfort, health, safety, and productivity.

Though the theory system of human factors is almost mature in the past years, there is no denying that novel terms generate all the day. Then how is the evolution of the subject? What is the hot topic these years? That is the purpose of our study. Most experts concentrate on the experiments research in their own field to solve actual problems, while our study aims to visualize the development and evolution of the subject of human factors and rich the outline due to the application of bibliometrics.

Bibliometrics is statistical analysis of written publications, such as books or articles. Bibliometric methods are frequently used to provide quantitative analysis of academic literature [1] or for evaluating budgetary spending [2]. Citation analysis is a commonly used bibliometric method, based on constructing the citation graph, a network or graph representation of the citations between documents. Many research fields use bibliometric methods to explore the impact of their field [3], the impact of a set of researchers, or the impact of a particular paper. Bibliometrics also has a wide range of other applications, such as in descriptive linguistics, the development of thesauri, and evaluation of reader usage. The purpose of this paper is twofold: to visualize the research to-date on human factors according to the intellectual framework and hot topics, and to propose directions for future research. The paper presents a comprehensive summary of previous research on the theme and identifies the most important issues. Based on the identified foundation of the research, key issues are highlighted, which should be addressed in future research.

2 Material and Method

2.1 Subject

The subject human factors originates from labor science in Europe. "Human Factors" created by HFES (Human Factors and Ergonomics Society) and "Ergonomics" created by IEA (International Ergonomics Association) in 1957 were the two journals published earliest and keep deep influence. Therefore, the literature is searched through SSCI. The authors search the title, abstract, keywords, and in some cases the full text. To ensure the authority of our research sample, proceedings, book reviews, editorials, letters are excluded from our search in Web of Science. Articles are the most common published materials suitable for our study, since they are easy to calculate valid bibliometric statistics, particularly in light of the heterogeneity of this type of materials.

2.2 Apparatus and Software

The primary goal of knowledge domain visualization is to detect and monitor the evolution of a knowledge domain [4]. CiteSpace II is to produce and analyze co-citation networks among highly cited articles. Developed by Chaomei Chen, CiteSpace II is a Java-based visualization application combining bibliometrics, information visualization methods, and data mining algorithms in an interactive visualization tool for the extraction of patterns in citation data. CiteSpace II supports nodes for author, article, journal, institution and country as well as for phrases extracted from titles and abstracts of articles. Chen et al. used the initial version of CiteSpace to reveal turning points in superstring revolutions in physics [5]. CiteSpace II is developed with new features related to three central concepts: burst detection, betweenness centrality, and heterogeneous networks. These concepts address three practical issues: identifying the nature of a research front, labeling a specialty, and detecting emerging trends and abrupt changes in a timely manner.

The general procedure of visualization analysis with CiteSpace II consists of nine steps: (1) knowledge domain identification, (2) data collection, (3) research front term extraction, (4) time slicing, (5) threshold selection, (6) pruning and merging, (7) layout styles selection, (8) visual inspection, and (9) pivotal point verification.

3 Procedure

3.1 Data Analysis

The authors retrieved 2959 records suited for our study published from 2000 to 2016. The full bibliographic records including authors, titles, abstracts and reference lists articles are downloaded. 14321 cited references were coved in our study. H-index of research on the knowledge domain is 76, indicating that 76 papers have been cited over 76 times. The top 10 related journals ranked by impact factors are analyzed. The most influential journal is about accident analysis and prevention, and other domains like human-computer interaction, industry, and behavior play a significant role in the rank. Human factors, ergonomics, and applied ergonomics list the top as comprehensive journals.

The field has seen some contributions from numerous disciplines, such as psychology, engineering, biomechanics, industrial design, behavior sciences, physiology, and anthropometry. In essence, the study of designing equipment, devices and processes fit the human body and its cognitive abilities. Human factors practitioners come from a variety of backgrounds, though predominantly they are psychologists (from the various subfields of industrial and organizational psychology, engineering psychology, cognitive psychology, perceptual psychology, applied psychology, and experimental psychology) and physiologists. Designers (industrial, interaction, and graphic), anthropologists, technical communication scholars and computer scientists also contribute. Different from the increasing trend in most research fields, the total number of articles in human factors volatility increase. The quantity of related articles slightly bursts in 2007 and 2014.

As for development of the subject in different countries (see, Table 1), research in the USA is of top of large quantity as well as good quality. Furthermore, the USA lead the cooperation among countries with high centrality. Though England create numerous articles, the portion of cooperated articles is small. Overall, cooperation in human factors research is far less than other categories.

Table 1. Ranking of the highly productive countries

Rank	Country	Centrality	Cooperation	Total papers	% of papers
1	USA	0.61	741	1301	44.0
2	England	0.10	49	370	12.5
3	Canada	0.11	70	264	8.9
4	Netherlands	0.00	24	179	6.0
5	Germany	0.13	55	169	5.7

3.2 Analysis of Co-citation

Reference analysis is one of the most important aspects of bibliometrics, allowing researchers to determine the influence of various authors and papers. Each node in Fig. 1 represents one cited document, and is depicted with a series of citation tree-rings across multiple time slices. The size of each node is proportional to the total co-citation frequency of the associated reference, while the thickness of a ring is proportional to the number of co-citations received in the corresponding time slice [6]. Each line connecting two nodes in the visualization represents one or more co-citation links involving the two references. The color of co-citation links shows the year in which the initial connection between the two documents. The authors selected top 24 cited papers from each time slice (1 year), and 332 nodes with 719 links are shown in the map. The density of the whole network is 0.0131.

In this visualization, red inner rings around a node signify a significant citation burst, indicating that citations of this document increased rapidly in a given time period. The size of the red rings represents the strength of this burst. As Fig. 1 indicates, the most frequently cited work in human factors (i.e., the node with purple trim and the largest citation rings) is a 2004 work edited by Lee [7]. The authors found that Lee's publication is the biggest (i.e., most impactful) core document in the timeline mode map shown in Fig. 3. For the purpose of managing imperfect automation, the review considers trust from the organizational, sociological, interpersonal, psychological, and neurological perspectives. It considers how the context, automation characteristics, and cognitive processes affect the appropriateness of trust. The work is widely applied to research on psychology, engineering industrial, computer science, and health information, etc. As for the reasons above, its centrality is also the highest in the whole network. Further will be discussed below. According to its citation history, Lee's work was highly cited until 2012. As for research on trust, it can be regarded classic still. Its total citation reached 410 according to SSCI, and 27 of them appear in Fig. 1.

The second most frequently cited document is a 1992 book edited by Wickens CD, which is made to understand and model human information processing and human performance. How these theoretical advances applied to improving human-machine interactions is discussed [8]. The book introduces engineering psychology. Perception, attention, cognition (both spatial and verbal) memory, learning and decision making, potential applications of these areas of cognitive psychology, control actions, time-sharing, error and stress, process control, complex systems and automation are all covered. In another word, it put into the broader framework of human factors and system design [9, 10]. Though the citation of the book nearly disappear since 2008, it

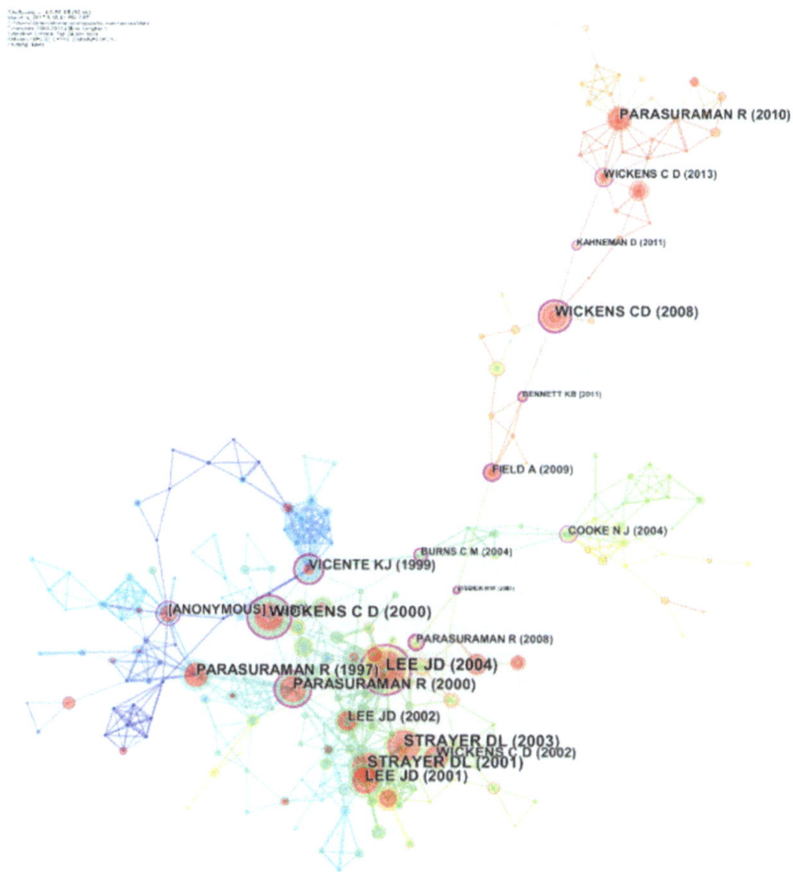

Fig. 1. Cited reference network for human factors

cannot be denied that the classic book lay the foundation of development in human factors. The two articles followed by Strayer DL are experiments on driver distraction. The two are the exploration about influence cell phone induced brings to simulated driving, using dual-task studies [11] and visual scene methodology [12]. Parasuraman R is expert in automation. They outline a model for types and levels of automation that provides a framework and an objective basis for deciding which system functions should be automated and to what extent [13]. Their earlier paper addresses theoretical, empirical, and analytical studies pertaining to human use, misuse, disuse, and abuse of automation technology [14, 15].

The top 10 cited papers of all years could be identified by noting the frequency with which they are cited (see, Table 2). Wicken CD, Lee JD, Strayer DL edit all the top 10 articles, and Parasuraman R, and Wickens CD edit most classical literature (papers with high half-life). Though effected by the two journals we select, they do make great contribution on human factors research. Parasuraman R's newly paper studies the different manifestations of complacency and automation bias in automation-induced phenomena, with attention playing a central role, and provide an integrated theoretical model [16].

Table 2. Top 10 articles with highest citation frequency

Citation	Author	Time	Title	Cluster	Half-life
27	Lee JD	2004	Trust in automation: designing for appropriate reliance	1	2
23	Wickens CD	2000	Engineering psychology and human performance	1	3
22	Strayer DL	2001	Driven to distraction: dual-task studies of simulated driving and conversing on a cellular telephone	0	3
22	Strayer DL	2003	Cell phone-induced failures of visual attention during simulated driving	0	1
19	Parasuraman R	2000	A model for types and levels of human interaction with automation	5	2
19	Lee JD	2001	Driver preferences for speech based interaction with in-vehicle systems	0	3
19	Parasuraman R	1997	Humans and automation: use, misuse, disuse, abuse	4	4
17	Wickens CD	2008	Multiple resources and mental workload	8	7
16	Lee JD	2002	Collision warning timing, driver distraction, and driver response to imminent rear-end collisions in a high-fidelity driving simulator	6	4
16	Wickens CD	2002	Multiple resources and performance prediction	15	6

Due to interest in applying the hierarchical cluster analysis as the main multivariate analysis technique and to the fact that this has proved particularly effective when an analysis like this is developed, a decision is made to not use Pearson's r correlation coefficient. A relative co-citation value or frequency was chosen instead: the Cocit-Score proposed by Gmür, which is used as a similarity measure to find a similarity-between-pairs matrix S, such that the similarity between a pair of papers (Pi, Pj) is given by normalizing their co-citation frequency with respect to the minimum and mean of the pair's respective citation counts:

$$S_{ij} = \frac{C_{ij}^2}{min(C_{ii}, C_{jj}) \times mean(C_{ii}, C_{jj})}. \quad (1)$$

The S_{ij} values are in the range [0, 1], with a larger value indicating a stronger association between a pair of papers. Note that C_{ii} on the main diagonal corresponds to the total number of citations for paper P_i.

As for the results obtained with the hierarchical conglomerates analysis, the application of that multivariate analysis technique to the matrix S resulting from the references co-citation analysis developed 68 clusters (including isolated points) in total, 9 of which have allowed us to check the existence (see, Table 3 and Fig. 2). Some of clusters link closely to one another.

Table 3. Prominent research clusters in human factors

Cluster	Size	Silhouette	Time
0	37	0.83	2002
1	32	0.858	2003
2	30	0.973	2006
3	26	0.951	2011
4	26	0.942	1996
5	16	0.918	2001
6	15	0.85	2002
7	15	0.925	1998
8	14	0.893	2009

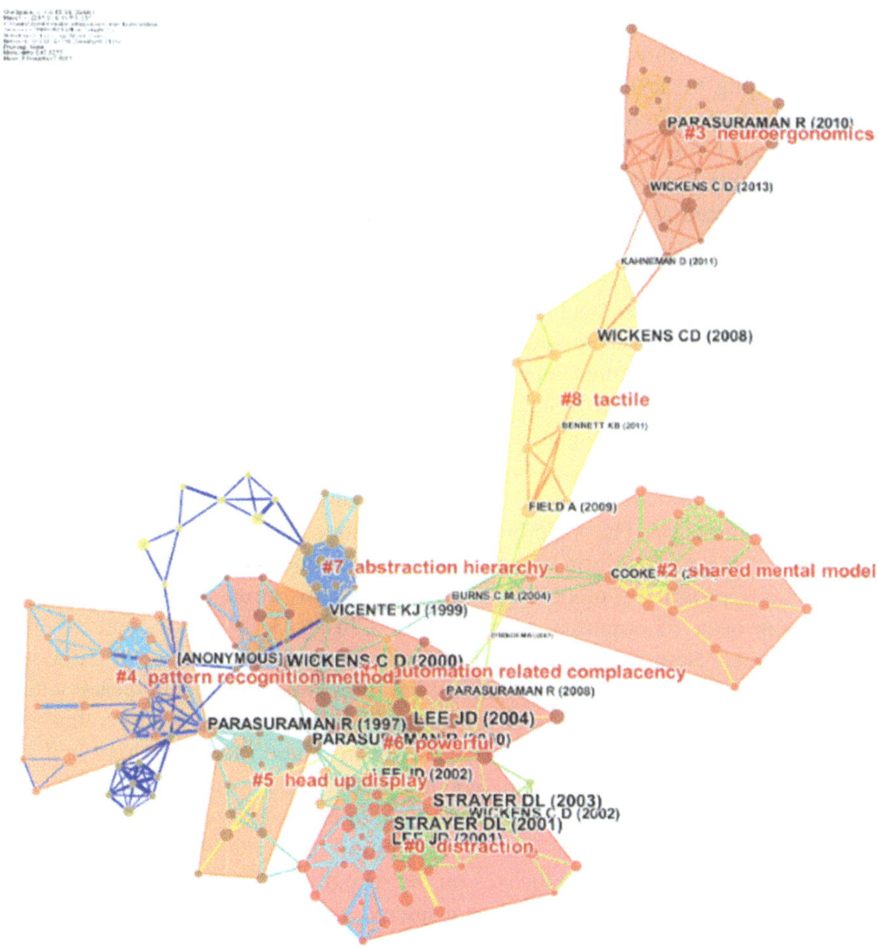

Fig. 2. Prominent research clusters in human factors

Most academic fields experience the evolution from theory basis to practice application and then to the methodology enhance, in comparison, research on human factors mainly based on experiments. According to the size of each cluster, the first five clusters are more evident. Nine subfields (clusters) were defined that constitute the intellectual structure and investigate their mutual relationships, including distraction, automation, factor analysis, warning, human-machine and other petite subfields which is isolated by the main system. The main findings of the co-citation analysis suggest that with the change of our focus in social phenomenon, the focus of human factors research changes from psychology to behavior, and the methodology tend to be mature and integrated. Cluster 0 explores driver distraction, belonging to the labor field. Algorithm development, distraction estimation, and visual search around traffic lights are main problems to solve. It appears around 2002. During the experiments, authors usually discuss age groups. The silhouette of the cluster is only 0.83, indicating there are some differences in the cluster. It takes up the core position in the whole network and links a lot with other groups, especially cluster 1, 5 and cluster 6. Plenty of research focus on distraction thus derived other problems or theory. It is of the active and practical meaning to develop the work in traffic safety administration. The emerging field of human factors in highway safety uses human factor principles to understand the actions and capabilities of road users - car and truck drivers, pedestrians, bicyclists, etc. - and use this knowledge to design roads and streets to reduce traffic collisions. Driver error is listed as a contributing factor in 44% of fatal collisions in the United States, so a topic of particular interest is how road users gather and process information about the road and its environment, and how to assist them to make the appropriate decision [17]. Cluster 1 is about controller attention, belonging to the human-machine relationship field. Increases in the use of automation have occurred across society, research has found that human operators often underutilize and overly rely on automated aids. In order to encourage people, especially youths to regard automation properly and correct their attitude, research in the cluster deals with measurements of complacency, personal computer use and other index. Compared with the two clusters mentioned above, the silhouette of Cluster 2 is much higher, which means its concentration is more clear. The topic is macro-cognition. Shared mental model is widely used. The term macro-cognition was created by Cacciabue and Hollnagel to distinguish the systemic approach of modelling cognitive systems from the traditional micro-cognitive approach of cognitive science [18]. Cluster 3 is newly founded named neuroergonomics. Neuroergonomics is the application of neuroscience to ergonomics. Research in the area of neuroergonomics has blossomed in recent years with the emergence of noninvasive techniques for monitoring human brain function that can be used to study various aspects of human behavior in relation to technology and work, including mental workload, visual attention, working memory, motor control, human-automation interaction, and adaptive automation. Cluster 4 is founded early around 1996. The topic is about pattern recognition method. Pattern recognition is a branch of machine learning that focuses on the recognition of patterns and regularities in data, although it is in some cases considered nearly synonymous with machine learning [19]. Cluster 5 takes account of head up display (HUD), situation awareness, performance and workload are discussed. HUD is any transparent display that presents data without requiring users to look away from their usual viewpoints. Although they were initially developed for military aviation, HUDs are now used in commercial aircraft, automobiles and other, mostly professional applications. As mentioned above, it links with cluster 0, indicating that the experiment methodology head up display can be also applied to the research or driver distraction. Cluster 6 is also linked closely to Cluster 0. Cluster 7 focuses on abstract hierarchy, and cluster 8 is on tactile display.

An alternative approach for viewing these clusters and their relationships is with timeline visualization (see, Fig. 3). The earliest paper appeared was published in 1993, one belongs to cluster 7, and the other two belongs to cluster 4, which reflects pattern recognition method and abstract hierarchy. Before 2000, cluster 4 and cluster 7 lead the mainstream, research then focus on psychology industry more. Fit between human and automation is significant. Cluster 1 and 5 almost start at the same time, related with each other closely. Between 2000 and 2005, as the traffic accidents become more and more serious, research on driver distraction appear. Papers belonging to cluster 0, 1, 2, 5 and 6 burst. Other clusters relate a lot to the two largest clusters. They provide situation for the research later. Cluster 3 and 8 are hot domains in recent years, indicating neuroergonomics and tactile display may be the frontiers of the subject.

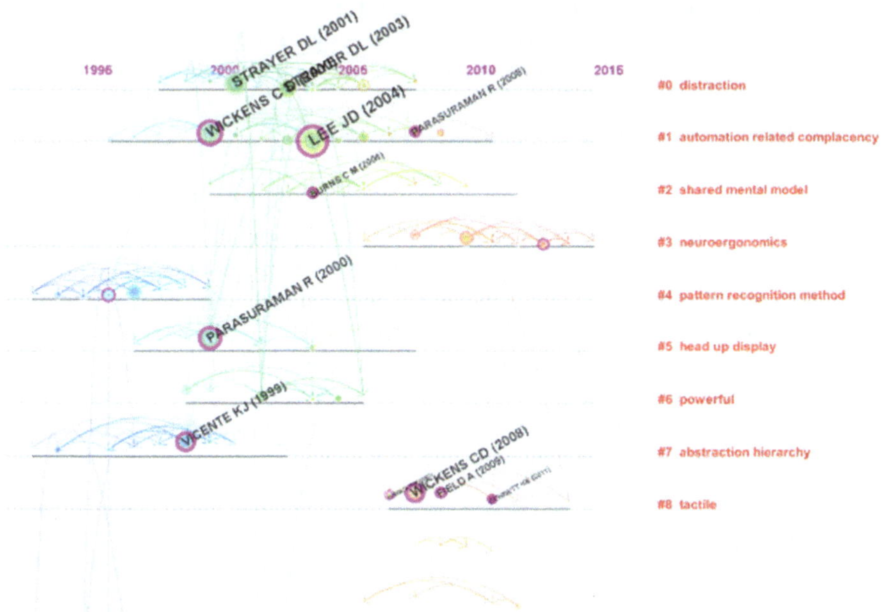

Fig. 3. Prominent research clusters with timeline visualization

The results also showed that the references which were cited most often were also the ones that have high centralities. Centrality is an indicator showing the ability of one reference to link with other references. Generally, references with high centrality are those which combine papers together in the same field.

The top 10 papers with highest centrality of all years can be identified (see, Table 4). Lee JD's work has the highest centrality, which means that this paper is

Table 4. Top 10 articles with highest centrality

Centrality	Author	Time	Title	Cluster	Half-life
0.37	Lee JD	2004	Trust in automation: designing for appropriate reliance	1	2
0.30	Vicente KJ	1999	Cognitive work analysis: towards safe, productive, and healthy computer-based work	7	1
0.27	Parasuraman R	2000	A model for types and levels of human interaction with automation	5	2
0.27	Field A	2009	Discovering statistics using SPSS	8	2
0.27	Eysenck MW	2007	Anxiety and cognitive performance: Attentional control theory	8	4
0.24	Wickens CD	2000	Engineering psychology and human performance	1	3
0.23	Parasuraman R	2008	Humans: still vital after all these years of automation	1	3
0.22	Wickens CD	2008	Multiple resources and mental workload	8	7
0.22	Bennett KB	2011	Bennett K. B., Flach J. M. (2011). Display and interface design: subtle science, exact art. Boca Raton, FL: CRC Press	8	4
0.21	Burns C M	2004	Ecological interface design	2	3

highly relevant to other papers in some sense [20]. In citation analysis, a node of high betweenness centrality is usually one that connects two or more large groups of nodes with the node itself in between, hence the term "betweenness" The network highlights nodes with high betweenness centrality with purple trim.

4 of the top 10 papers are from cluster 1 and 5, and the other 4 are from cluster 8. 4 papers have been mentioned above. The key articles published earlier provide basic model and methodology for experiments [21–24]. Though research on tactile displays begins early, it processes great potentials and prospects.

4 Conclusions

This article uses a generic approach to detecting and visualizing emerging trends and transient patterns in scientific literature. The work makes substantial theoretical and methodological contributions to progressive knowledge domain visualization. The latest literature on human factors or ergonomics from 2000 to 2016 are studied. In use of bibliometric analysis, the authors draw conclusions as follows. Far different from other subjects, research on human factors own its unique regulations. Concentrated on empirical and experiments, the subject lacks of academic cooperation. With the foundation during World War II, the USA keeps the lead the academic achievements in diverse countries. The fit of humans, machines and environment is the main mission of the subject. Distraction and automation are topics discussed most, and variety of advanced methodology generated under the situation. In addition, 9 main domains in human factors are explored through visualization. It's credible that research on neuroergonomics and tactile display may be the frontiers of the subject.

Acknowledgment. This research was supported by the National Natural Science Foundation of China (No. 71371025).

References

1. Nicolaisen, J.: Bibliometrics and citation analysis: from the science citation index to cybermetrics. J. Assoc. Inf. Sci. Technol. **61**(1), 205–207 (2010)
2. Coughlin, D.M., Campbell, M.C., Jansen, B.J.: A web analytics approach for appraising electronic resources in academic libraries. J. Assoc. Inf. Sci. Technol. **67**(3), 518–534 (2016)
3. Pilkington, A., Meredith, J.: The evolution of the intellectual structure of operations management, 1980–2006: a citation/co-citation analysis. J. Oper. Manag. **27**(3), 185–202 (2009)
4. Chen, C.: Searching for intellectual turning points: progressive knowledge domain visualization. Proc. Natl. Acad. Sci. **101**(1), 5303–5310 (2004)
5. Chen, C., Hu, Z., Liu, S., Tseng, H.: Emerging trends in regenerative medicine: a scientometric analysis in CiteSpace. Expert Opin. Biol. Ther. **12**(5), 593–608 (2012)
6. Chen, C.M., Kuljis, J.: The rising landscape: a visual exploration of superstring revolutions in physics. J. Am. Soc. Inf. Sci. Technol. **54**(5), 435–446 (2009)
7. Lee, J.D., See, K.A.: Trust in automation: designing for appropriate reliance. Hum. Factors J. Hum. Factors Ergon. Soc. **46**(1), 50–80 (2004)
8. Wickens, C.D.: Engineering Psychology and Human Performance. HarperCollins Publishers, New York (1992)
9. Wickens, C.D.: Multiple resources and mental workload. Hum. Factors J. Hum. Factors Ergon. Soc. **50**(3), 449 (2008)
10. Lee, J.D., Mcgehee, D.V., Brown, T.L., Reyes, M.L.: Collision warning timing, driver distraction, and driver response to imminent rear-end collisions in a high-fidelity driving simulator. Hum. Factors J. Hum. Factors Ergon. Soc. **44**(2), 314 (2002)
11. Strayer, D.L., Johnston, W.A.: Driven to distraction: dual-task studies of simulated driving and conversing on a cellular telephone. Psychol. Sci. **12**(6), 462–466 (2001)
12. Strayer, D.L., Drews, F.A., Johnston, W.A.: Cell phone-induced failures of visual attention during simulated driving. J. Exp. Psychol. Appl. **9**(1), 23 (2003)
13. Parasuraman, R., Sheridan, T.B., Wickens, C.D.: A model for types and levels of human interaction with automation. IEEE Trans. Syst. Man Cybern. - Part A Syst. Hum. **30**(3), 286–297 (2000)
14. Parasuraman, R., Riley, V.: Humans and automation: use, misuse, disuse, abuse. Hum. Factors J. Hum. Factors Ergon. Soc. **39**(2), 230–253 (1997)
15. Parasuraman, R., Wickens, C.D.: Humans: still vital after all these years of automation. Hum. Factors J. Hum. Factors Ergon. Soc. **50**(3), 511 (2008)
16. Parasuraman, R., Manzey, D.H.: Complacency and bias in human use of automation: an attentional integration. Hum. Factors J. Hum. Factors Ergon. Soc. **52**(3), 381 (2010)

17. Campbell, J.L., Lichty, M.G., et al.: National Cooperative Highway Research Project Report 600: Human Factors Guidelines for Road Systems, 2nd edn. Transportation Research Board, Washington, D.C. (2012)
18. Cacciabue, P.C., Hollnagel, E.: Simulation of cognition: applications. In: Hoc, J.M., Cacciabue, P.C., Hollnagel, E. (eds.) Expertise and Technology: Cognition and Human-Computer Cooperation, pp. 55–73. Lawrence Erlbaum Associates, Mahwah (1995)
19. Bishop, C.M.: Pattern Recognition and Machine Learning. Springer, New York (2006)
20. Dimoka, A., Banker, R.D., Benbasat, I., Davis, F.D., et al.: On the use of neurophysiological tools in is research: developing a research agenda for NeuroIS. MIS Q. **36**(3), 679–702 (1989)
21. Wickens, C.D., Hollands, J.G.: Engineering Psychology and Human Performance, 3rd edn. Prentice-Hall, Upper Saddle River (2000)
22. Wickens, C.D.: Multiple resources and performance prediction. Theor. Issues Ergon. Sci. **3**, 159–177 (2002)
23. Aldridge, L.C., Lansdown, T.C.: Driver preferences for speech based interaction with in-vehicle systems. Hum. Factors Ergon. Soc. Annu. Meet. Proc. **43**(18), 977–981 (1999)
24. Vicente, K.: Cognitive Work Analysis: Towards Safe, Productive, and Healthy Computer-Based Work. Erlbaum, Mahwah (1999)

Building the Knowledge-Based System of Machining Supplier Matching

Shang-Liang Chen and Hoai-Nam Dinh(✉)

Institute of Manufacturing Information and System,
National Cheng Kung University, Tainan, Taiwan
slchen@mail.ncku.edu.tw, dinhhoainam52@gmail.com

Abstract. Customer orientation is one of the key factors in consumer centric situation. The pervious methods consist of processing suppliers, such as telephone interviews or official website browsers. These are incapable of providing immediate and efficient manufacturing information. This paper builds a system, which gets the information processing, and guidelines for supplier selection process. The machining knowledge database collects the relevant information through manufacturers and customer orders, such as monitoring possible customer behavior analysis, behavioral analysis of vendor, and platform profitability analysis of system. Accordance with users' needs, the system recommend suitable processing providers.

Keywords: Machining knowledge database · Production Cost · Selecting Supplier

1 Introduction

Consumers can directly or indirectly find information in the enterprise-marketing platform to buy the production they need. They can also respond to C2M (Customer to Manufactory) trend of customer-oriented strategy. This is a target machine equipment manufacturing enterprise. Many researches have verified market-oriented enterprises eligible for interest rates, both the sales growth rate and product development success are connected [1, 2]. Narver et al. [3] has separated into market-oriented proactive market orientation and reactive market orientation. Supply chain concept matures, changing the relationship between business and suppliers, from the previous trading simple supply and demand into mutually beneficial cooperation and sharing resources [4]. Reduced data leakage and data security can be improved by the protection from the firewall: an internal private cloud. Mixed cloud is a combination of the two mentioned above advantages, it also has the flexibility of public cloud infrastructure and security advantage of a private cloud data [5]. Cloud service modules assist machine manufacturing enterprises by monitoring information. Business can choose according to their plan through the public cloud needed to build private cloud data. Based on the enterprise collaboration platform, Chebrolu [6] proposed a hybrid cloud architect. The private cloud planning security mechanisms consist of identity authentication and classification authority. The public cloud service section consist of web server, application server, and database, which provide the required service and set up for a

T. Ahram and W. Karwowski (eds.), *Advances in Human Factors, Software, and Systems Engineering*, Advances in Intelligent Systems and Computing 598,
DOI 10.1007/978-3-319-60011-6_5

multi-layer firewall to enhance security. Ni-Di and Ming-Xian [7] applied AHP (Analytic Hierarchy Process) to agricultural food. The four covered criteria for this supplier selection was quality, delivery, service and cost. The results revealed that the highest criteria are the quality and the second highest is the cost in the business environment. To create a new virtual enterprise in partner selections, AHP and fuzzy method was combined by Jain Cao et al. [8]. Based on the theory of fuzzy AHP each attribute could not be given. Because of this, it is recommended for partners. The current market has a lack of integration for supplier information processing platform and have recommended mechanism. To improve the above problems, it can seek to enhance the efficiency of processing providers, and thus enhance the competitiveness of the production of machine equipment manufacturing enterprises.

2 Methods

2.1 Structure Design of Intelligent System for Machine Tool

The study proposes a system which enables the user to obtain information processing suppliers, and quick response manufacturing capacities, using responsive web design (RWD) to develop cloud services module. Its advantage can be the development to support a variety of interfaces to use, and change the page layout and typography at different resolutions. Therefore, the user can take advantage of the device can be networked computer browser or mobile device, etc. The establishment of a high efficiency can be achieved searching and filtering with process flow, which is shown in Fig. 1.

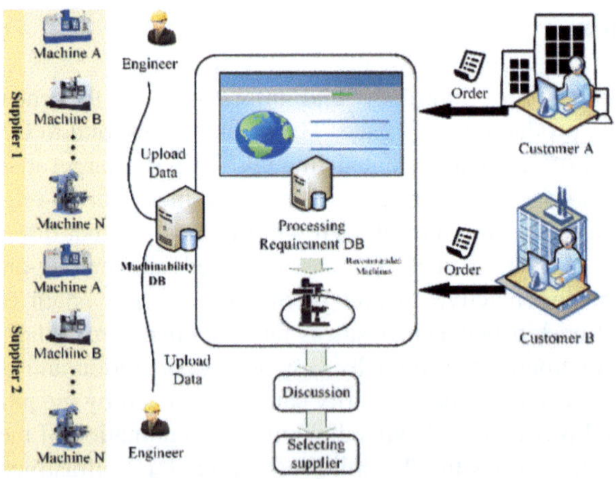

Fig. 1. Overall the system for machine tool

From orders of customer allows us to estimate and simulate the processing components through the Siemens PLM NX software. The use of quality, delivery and price analysis as levels of assessment criteria, quality criteria section. It will evaluation the criteria for quality evaluation and becomes more defined each of the criteria described below:

(1) Evaluation: contains supplier produced the work quality and service quality suppliers.
(2) Delivery: refers to the degree of precision of the supplier to deliver the goods.
(3) Price: refers to the consumer and the supplier for paying and purchasing the productions.

2.2 Knowledge Representation of Machine Process Selection

Previous researches have proposed some solutions to identify machine features and different geometric shapes with topological properties. The form features include geometric shapes and manufacturing operations. Data features can be identified from surface or volumes. The production cost is calculated by machining features in [9]. Work-piece machining cost is one of the critical factors, Shehab and Abdalla [10] researched plastic injection molding products that calculated the cost of a set of intelligent knowledge systems. Through the process of supplier data collection and platform promotion, the processing database information of the Cloud Database Center is continually updated to provide users with the latest and most complete information. The processing is shown in Fig. 2.

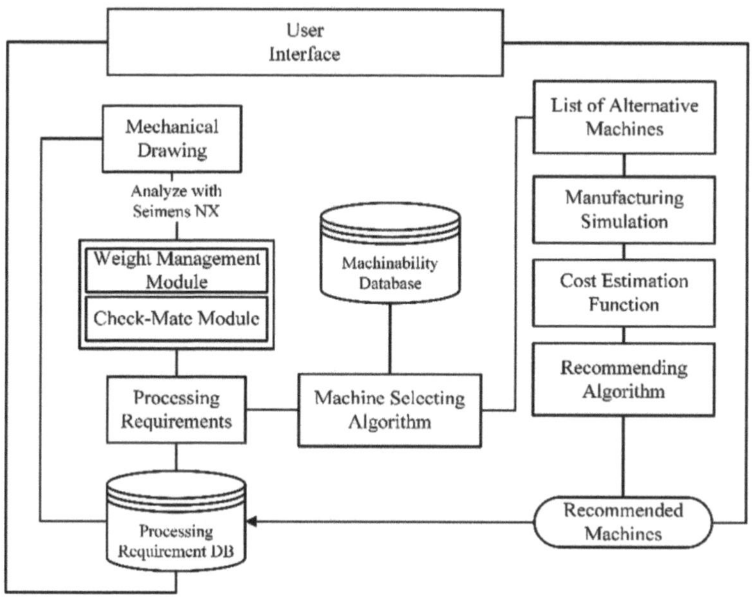

Fig. 2. The flowchart of system

Database is designed as an ER model, is shown in Fig. 3. The supplier through processing throughout data collection and platform to promote the process of cloud database vendor data centers constantly updates to provide users with the latest and most complete information. This achieve a collection of multiple use purposes.

Management module provides manufacturing customers and suppliers to use process, which includes the following functional modules:

(1) Basic data maintenance and views: You can view this after user login and modify your personal information and password.
(2) Personal Order Management: Provides consumers their own orders and then confirm the order by the transaction list. Consumers also check or modify the order information.

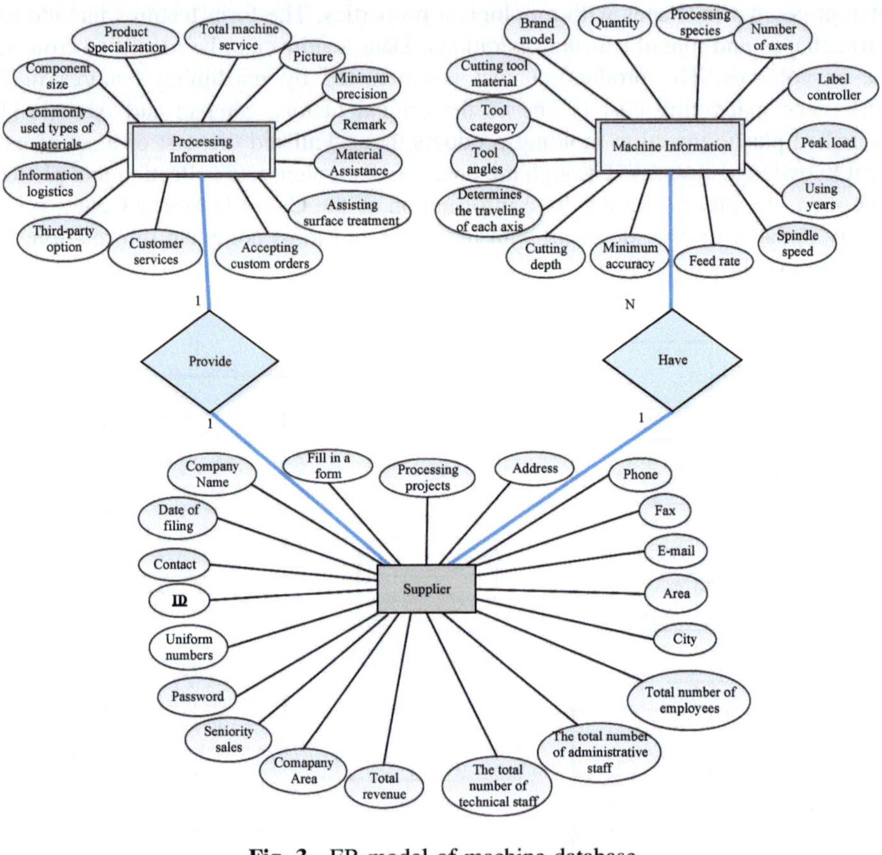

Fig. 3. ER model of machine database

2.3 On-Line Matchmaking Process

For drawing parts has high confidentiality of the clients, online matchmaking of this project will enable customers to take advantage of convenient search system to find the appropriate recommendation of manufacturers to provide processing services, without the need to open a private drawing. According to the drawing processing, customer via search system of the present program and their needs were provided tag cloud screening system quickly filtered by the back-end database to meet all the labels of the

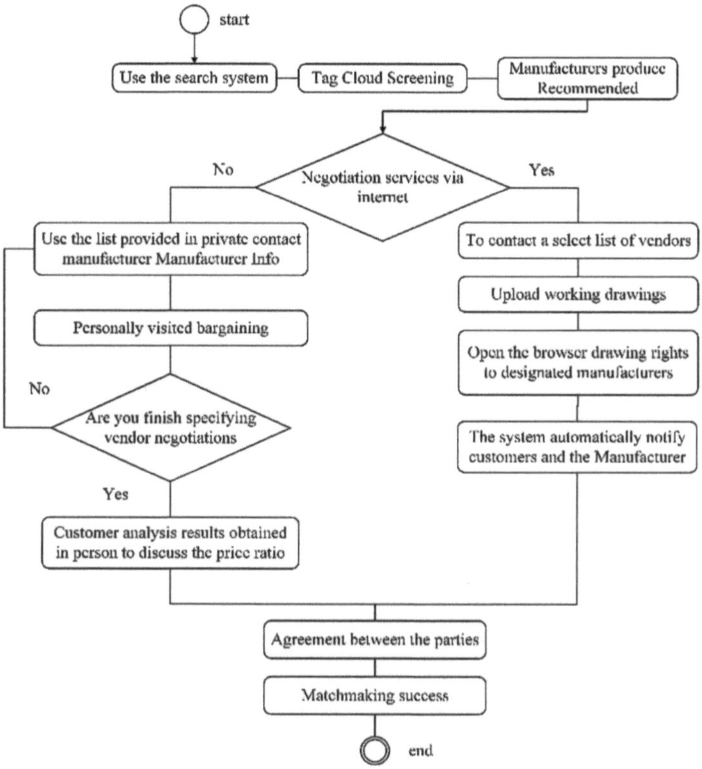

Fig. 4. Flow chart for on-line matchmaking process planning

third-party processing. All this data is collected and shows the recommended list to the client. In this case the customer may in turn decide to contact the firms themselves or use the services of this program's rapid negotiation discussions with several manufacturers which allows them to simultaneously negotiated and completed matchmaking. Matchmaking initiative to search the flow chart shown in Fig. 4.

2.4 Planning Hierarchy

In the next phase, planning hierarchy, the user inputs the workpiece diagram information which is provided by the conditions in this study. The machine filter criteria, offers three common conditions for the service suppliers aiding users filter out matches. AHP three-level architecture diagram shown in Fig. 5. With the rapid evolution of manufacturing technology, machining tools have been able to reach into the high-speed, high-precision cutting stage, however, it is a comprehensive high-precision cutting machine design, high-speed spindle design, materials technology, control technology and thermal compensation technology; a number of emerging technologies evolved.

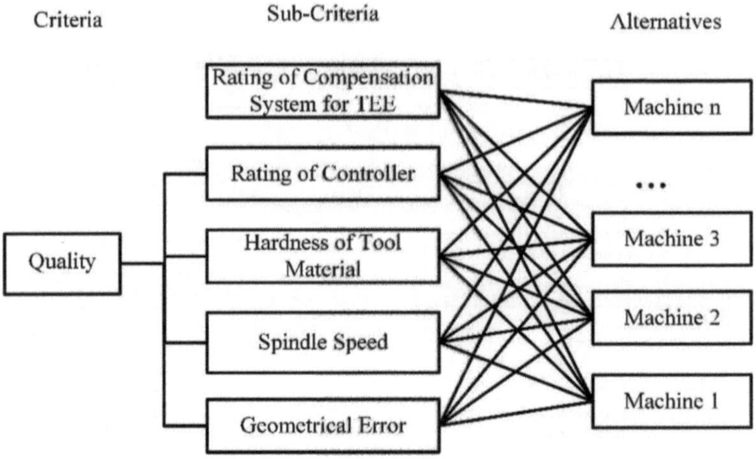

Fig. 5. Hierarchy diagram of the processing quality

3 Case Study

In this case study, engineering drawing is in Fig. 6. We get the work-piece of geometry conditions by analyzing NX in Table 1.

There were 5 manufacturers in this case study, the machine information is listed in Table 2. After the processing condition in workpiece, the features of object are converted to SQL language. After clicking the filter button, the module will filter the machine through the restriction condition string to the machine data table, and the result is shown in the DataGridView, as Fig. 7. Finally, the system recommend for client two machine tool which were suitable in Table 2 include in machine numbers 3 and 5.

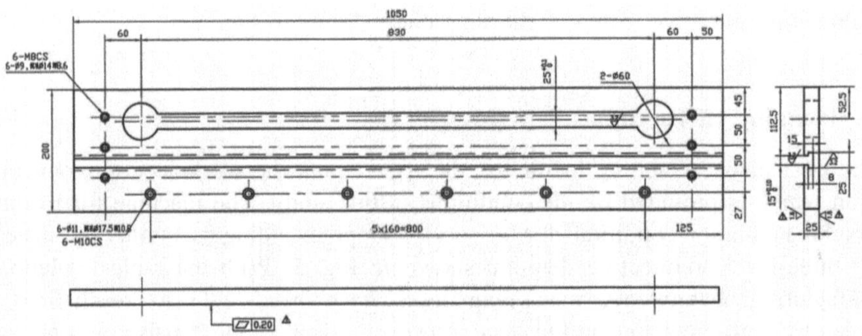

Fig. 6. Drawing of workpiece

Table 1. Geometric condition of workpiece

Geometric	Unit
Length (mm)	1050
Height (mm)	25
Material	Q235A
Density	7.85
Cutting speed (MPM)	110
Accuracy (CNS)	IT10
Width (mm)	200
Volume (mm^3)	$5.25 * 10^6$
Hardness (BHN)	165
Mass (Kg)	41.21
Spindle speed (rpm)	3503

Table 2. Information of machine tool of manufacturers

ID	Brand	Model	Length (mm)	Width (mm)	Height (mm)	Weight (Kg)	Spindle speed (rpm)	Accuracy (CNS)	Status
1	Y***	F**6A	700	420	560	300	10000	11	0
2	Y***	F**5A	1000	520	690	500	10000	9	1
3	Y***	**F**2A**	**1120**	**520**	**690**	**500**	**10000**	**7**	**1**
4	O***	V**55	460	460	610	300	8000	10	0
5	O***	**V**5C**	**1100**	**510**	**710**	**750**	**6000**	**8**	**1**

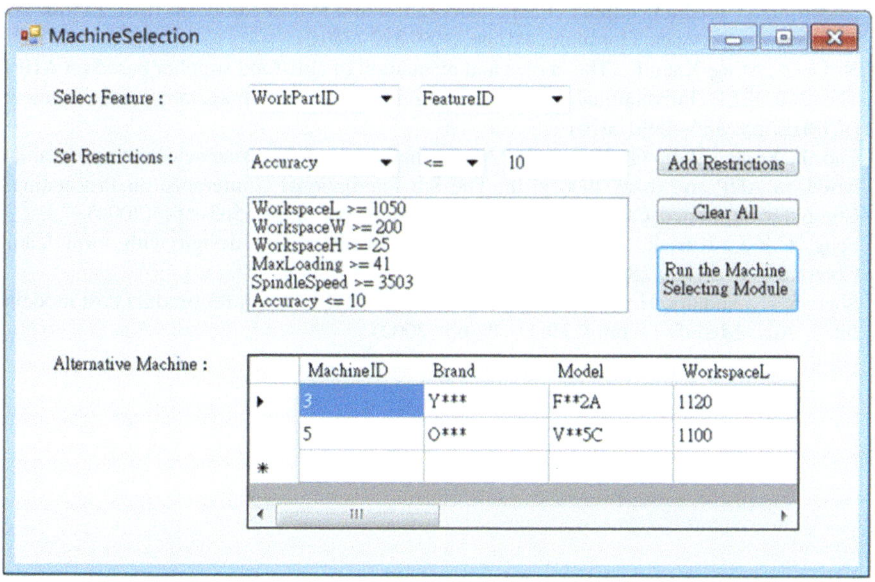

Fig. 7. Recommend the machine tool in machine selecting module

4 Conclusions

This paper proposed the design the knowledge-based system, which offers a flexible matching service of subcontract work for client and vendor. Processing the cloud database replaces the traditional paper processing needs. The user via the drawing information can find the type of machining, the workpiece size, workpiece material, workpiece precision and surface roughness, are set in accordance with the work requirements. Matchmaking mechanism will search the database that matches the conditions of the machine filter out and finds these conditions have machine suppliers. The traditional model can be replaced through achieving the processing cost estimation to gauge the products cost. Allowing the user to seek situations and the user's clicked suppliers. The type of processing offers the type of user needs via the search history.

References

1. Atuahene-Gima, K.: An exploratory analysis of the impact of market orientation on new product performance. J. Prod. Innov. Manag. **12**(4), 275–293 (1995)
2. Deshpande, R., Farley, J., Webster, F.: Corporate culture, customer orientation, and innovativeness in Japanese firms: a quadratic analysis. J. Mark. **57**(1), 23–27 (1993)
3. Narver, J., Slater, S., MacLachlan, D.: Responsive and proactive market orientation and new product success. J. Prod. Innov. Manag. **21**(5), 334–347 (2004)
4. Kemppainen, K., Vepsalainen, A.: Trends in industrial supply chains and networks. Int. J. Phys. Distrib. Logist. Manag. **33**(8), 701–719 (2003)
5. Rajan, S., Jairath, A.: Cloud computing: the fifth generation of computing. In: International Conference CSNT, pp. 665–667 (2011)
6. Chebrolu, S.: Enabling supply chain collaboration in a hybrid cloud. In: IEEE Eighth World Congress on Services, Honolulu, HI, pp. 309–312 (2012)
7. Ni-Di, Z., Ming-Xian, L.: The choice and evaluation of agri-food supplier based on AHP. In: The 2nd IEEE International Conference on Information Management and Engineering (ICIME), pp. 484–489 (2010)
8. Cao, J., Ye, F., Zhou, G., Huang, H.: A new method for VE partner selection and evaluation based on AHP and fuzzy theory. In: The 8th International Conference on Proceedings of Computer Supported Cooperative Work in Design, vol. 2, pp. 563–566 (2004)
9. Feng, C.X.J., Kusiak, A., Huang, C.C.: Cost evaluation in design with form features. Comput. Aided Des. **28**(11), 879–885 (1996)
10. Shehab, E., Abdalla, H.: An intelligent knowledge-based system for product cost modelling. Int. J. Adv. Manuf. Technol. **19**(1), 49–65 (2002)

Software Cost Estimation for User-Centered Mobile App Development in Large Enterprises

Maria Lusky(✉), Christoph Powilat, and Stephan Böhm

Center of Advanced E-Business Studies (CAEBUS),
RheinMain University of Applied Sciences,
Unter den Eichen 5, 65195 Wiesbaden, Germany
{lusky, powilat, boehm}@caebus.de

Abstract. Since development processes for mobile applications (apps) are becoming more user centered and agile, effort and cost estimation for app development projects in large enterprises faces new challenges. In this paper, we propose a new experience-driven approach for effort and cost estimation in software development projects. A Delphi study is conducted that takes into account different perspectives in mobile app development. Recurring app features are identified and associated with effort estimates and their variations based on different roles, perspectives, and complexity levels. In order to utilize our findings, a prototypical tool is introduced that allows effort and cost estimation based on incomplete information at an early stage in a project.

Keywords: Human Factors · User centered design · Software cost estimation · Software effort estimation · Mobile application development · Systems Engineering

1 Introduction

Mobile application (app) development has to cope with a dynamic environment and the need for early user involvement to create good user experience (UX). For this reason, even large enterprises (e.g., IT service providers) abandon the traditional waterfall model and introduce user-centered design [1] as well as agile software development processes [2]. In line with this, cost estimation becomes more challenging. Existing software cost estimation approaches [3–5] are not sufficient, since they are too cumbersome for the calculation of agile and lean projects [6]. They often rely on detailed requirement specifications, while agile and user-centered projects typically start with basic prototypes refined within an iterative process. As a result, efforts for developing an app can only be estimated vaguely. In this situation, expert estimations derived from previous projects and experience are often used instead, but are hardly transparent for non-experts [7]. While small enterprises (e.g., start-ups) might be able to cope better with this uncertainty, due to a more direct communication and short decision-making paths, larger enterprises need a more structured and reliable process to plan and allocate resources. Therefore, mobile app engineering – especially in larger enterprises – requires new methods for cost and effort estimation independent from a specification-based calculation logic. Thus, it is necessary to convert ex-ante evaluation of costs and efforts

T. Ahram and W. Karwowski (eds.), *Advances in Human Factors, Software, and Systems Engineering*, Advances in Intelligent Systems and Computing 598,
DOI 10.1007/978-3-319-60011-6_6

into an iterative cost estimation process along the collection of user feedback. Moreover, this approach has to facilitate exchanging and storing experiences on development efforts and costs in a knowledge base.

Against this background, this paper suggests an approach to support cost estimation for app development based on the identification and evaluation of software artifacts. These artifacts represent app features, e.g., functional components, that are frequently subject to implementation in development projects of a specific enterprise [8]. The software artifacts are building the foundation of a knowledge management tool that enables project managers to derive a rough cost estimate for an app project at an early stage. Accuracy of the estimates can be improved by enabling project managers to adjust this information as the project proceeds and by accumulating cost estimates across completed projects. A Delphi study within a large German IT service provider is conducted to demonstrate the basic feasibility of such an experience-driven approach. The study takes into account the perspectives and experience of both developers and user experience designers. A first round of this survey comprises the identification and classification of software artifacts. In subsequent rounds, these artifacts are evaluated with regard to development efforts and cost drivers. As a result, a set of recurring software artifacts as well as the associated effort estimates and their variations based on different roles, perspectives, and complexity levels will be presented.

Following this introduction, Sect. 2 presents the status quo and related work on effort estimation approaches in software development. Section 3 introduces our Delphi study, while Sect. 4 shows the results of our analysis. In Sect. 5, key findings and needs for further research are discussed. Building on that, Sect. 6 focuses on managerial implications for the implementation of our approach in practice. Section 7 emphasizes on key insights as well as limitations of our approach and concludes this paper.

2 Related Work

Software development effort and cost estimation has been widely studied to this day. By 2007, Jorgensen and Shepperd [9] had already identified more than 300 studies in this field. Yet, despite the general interest in this subject and the growing importance and rapid proliferation of mobile technologies and services, very few studies investigate effort and cost estimation with regard to the development of mobile applications. In fact, we have found only 15 studies that explore methods to estimate the size, efforts or costs of mobile app development projects.

2.1 Cost and Effort Estimation in Mobile App Development

In 2012, Schmietendorf [10] published a positional paper on the specific characteristics of mobile app development and the resulting challenges and issues for effort and cost estimation. In addition, the author presents alternative approaches to estimate the functional size of mobile applications as well as the efforts and the costs of such projects. However, the applicability of these methods is not further examined in this paper.

Most of the subsequent research investigates the suitability of various existing software metrics to calculate and estimate the functional size of mobile applications. For example, [11] adapts the COSMIC Function Points approach to measure the size of mobile apps based on its functional user requirements. [12] examine the appropriateness of traditional code-based software metrics (e.g., Lines of Code) for source code of mobile apps. [4] further analyse the use of COSMIC Function Points to measure the size of mobile apps and investigate whether the results correlate with code-based measurements.

Very little research has been conducted on how to estimate efforts and costs of mobile app development. [13] propose a new approach to predict the effort to develop mobile apps. First, the functional size of the app is estimated based on the Functional Size Measurement of the Finnish Software Measurement Association (FiSMA FSM). In the next step, the estimator has to evaluate a set of productivity factors (e.g., complexity of functional requirements) and product specific characteristics (e.g., level of offline capabilities). [6] present an analogy-based approach in which the functional size of the app is measured based on its screens and the containing function points. The method also includes the evaluation of various scaling factors (e.g., architecture and risk treatment) and cost drivers (e.g., technological maturity of the platform) to take into account the constraints and specific requirements of mobile app development. [14] investigate the value of various functional and code-based measures in order to estimate the effort needed to develop a mobile application and the number of its graphical components.

However, it is unclear if the presented approaches are suitable for effort and cost estimation in mobile app development. While the proposals by [13] and [6] represent guidelines that lack empirical validation or verifying use cases, the estimation method by [14] is based on specific user requirements that are usually not available until later stages in the development process of mobile applications.

2.2 Web Tools for Cost and Effort Estimation

While empirically validated estimation methods are still lacking, numerous websites provide a more practical orientated cost estimation for mobile app development. The calculation is mostly based on software components ("app features") that each provide a specific functionality within mobile applications (e.g., login, location tracking, push notifications) and/or several scaling or influence factors (e.g., number of platforms), which determine the functional size of the app. In order to get an estimate of the development costs, the user has to either (1) select the respective app features [15], (2) answer questions that aim to determine the functional size of the application [16] and/or (3) provide information on the quantity of specific app components (e.g., number of screens) [17].

Most tools, such as the app estimator by [18], represent a combination of those different approaches and take into account various company-specific cost factors (e.g., hourly wages for developers) and the estimated amount of effort (i.e., the number of work hours) needed to develop the respective mobile application to calculate the cost of development. Such tools provide a level of abstraction between the estimator and the underlying calculation logic and hence allow a simple and intuitive usage that enables even non-experts to conduct a cost estimation for mobile apps. In addition, no detailed knowledge on the functional size of the proposed application is required – thus, such calculation can be performed based on vague information at an early stage of the development process. However, those web calculation tools can not be directly used for effort and cost estimation in large enterprises. This is because they are typically aligned to the cost structure of a specific app developing company and provided with a focus on marketing and customer acquisition. For example, it can be assumed that the portfolio of features and costs presented in such calculation tools is aligned to attract specific target groups and based on combined costing rather than reflecting artifacts with a high level of experience or reuse within the developing company. In addition, the presented tools do not reveal the methodology used or data origin and thus lack scientific foundation.

2.3 Other Tools for Cost and Effort Estimation

The Constructive Cost Model (COCOMO) is another popular model and tool for effort and cost estimation in software development and was first specified and published by [19] in 1981. Since then, the method has been revised many times to meet the changing requirements of software development. In each version, the size of the software is measured in Kilo Delivered Source Instructions (KDSI). In the basic version (COCOMO 81), the effort in man-months is calculated based on the number of KDSI and two constants representing the overall complexity of the project. The intermediate model (Intermediate COCOMO) further includes the evaluation of fifteen effort and cost drivers (e.g., required software reliability) to take into account the specific project characteristics. In 1995, [20] published COCOMO II to address emerging iterative and reuse driven software development approaches. This model includes three sub-models (i.e., application composition, early design, and post-architecture) that vary in complexity with regard to the stage of the development process.

However, all versions of COCOMO are general-purpose models that do not take into account the product specific characteristics and dynamic environment of mobile app development. Accordingly, the application of any version might lead to inaccurate and imprecise estimates in this field.

3 Approach and Evaluation

Against this background, our approach builds on the principles of the tools discussed in Sect. 2.2 and takes into account the specific conditions that characterize the development process of mobile applications in large enterprises. An evaluation of the core principles of this approach based on a Delphi study forms the scientific foundation for the conceptualization and prototypical implementation of a tool that aims to support cost and effort estimation in mobile application development projects in large enterprises. In order to acquire the foundation and investigate the basic feasibility of such an approach, the following research objectives have been defined:

(1) Can expert knowledge and experience of app developers and UX designers be used to identify reusable artifacts and cost/effort drivers?
(2) Does the localization of effort between conceptualization and programming vary among the different features?
(3) Can a two-step process be used to collectively estimate costs/efforts for frequently used artifacts (identified in the first step)?

It should be noted here, that this Delphi approach is just a proxy for a much more sophisticated future software cost estimate process. The long-term objective of our project is to develop a framework for artifact reuse and cost estimates based on an enterprise knowledge base that is continuously enhanced with project data. The evaluation of our approach here was conducted to gain some first insights in the feasibility and workability of such an approach. Would developers in large enterprises – used to work with sophisticated software cost estimate tools (i.e., function points) – take part in a more collaborative and interactive estimation process and are they willing to provide their expert knowledge with the required high level of abstraction? This is why the focus of our evaluation was more on the appropriate level of detail and to get some insights on the granularity of information that could be derived from the developer and UX designers. At this stage of the study it was not intended to get reliable or generalizable cost and effort estimates for the artifacts. This as well as the accuracy of the estimates that can be derived by our approach are subject to further work.

3.1 Methodology

The evaluation was conducted as a practice-oriented variation of a Delphi study. In general, a Delphi study is a systematically multistage survey during which a group of experts assesses a matter at hand [21]. In the context of this work, persons that work in mobile app development in large enterprises and therefore have the required thematic experience are considered experts. For the second round, the Program Evaluation Review Technique (PERT) was used, a weighted calculation method that can generally be used for increasing the accuracy of effort estimation. For determining the decisive estimate (A in Formula 1), the following formula is used:

$$A = \frac{bc + 4 * lc + wc}{6}. \tag{1}$$

Following this method, an expert makes three estimations for development effort that are optimistic (best case, bc), realistic (likely case, lc) and pessimistic (worst case, wc) [22].

3.2 Procedure

The participants were asked about different partial matters derived from our research objectives in sequential survey rounds. Thus, the procedure was as follows:

(1) A set of app features was identified in a prestudy by evaluating several mobile apps of a large German IT service provider with more than 3,500 employees. This set of features was validated by existing feature sets (from the web tools) and by experts from the field.
(2) In a first round of the survey, those features that are frequently used in mobile apps of the enterprise were identified as standard features. Furthermore, the degree of customization as well as cost drivers were specified.
(3) Building on that, in a second round of the survey an effort (man-days) estimation of the standard features was made, taking into account the different perspectives of conceptualization (UX design) on the one hand and programming on the other hand.

For the prestudy, five apps of the IT service provider focusing on logistics and transportation were compared regarding their features. Additionally, several features were taken from online calculation tools for app development. Thereof, a list comprising 80 standard app features organized in 17 categories was derived. This list was still too long to be distributed for more detailed feedback within the enterprise. This is why, with the help of corporate experts, this list was reduced to 16 app features by summarizing similar ones and sorting out irrelevant features. The 16 features that served as a basis for the first survey round of our study were: registration and login, chronological lists, search, push notifications, calendar and time, hardware access, reservation and booking, file upload, favorites, map view, localization, scanning feature for optical codes, comment feature, messaging, sharing, navigation. In the second round, the list was reduced to the ten most frequently implemented features. The survey was conducted using a standard online survey tool. Besides demographic information, such as gender and age, the participants were asked about their professional experience, comprising the field of activity in mobile app development, project experience and the share of agile methods in these projects.

In the first round of the survey, the participants were asked three questions for each feature: (1 – Perceived Frequency, PFQ) How often is the app feature implemented in mobile apps in your company (1 = never, 5 = everytime)? (2 – Degree of Customization, DOC) To which degree is the feature usually needed to be adapted to customer requirements (1 = very little, 5 = very strongly)? (3 – Location of Effort, LOE) Where do you locate the main effort for developing the feature (1 = exclusively in conceptualization, 5 = exclusively in programming)?

In the second round of the survey, the participants were asked to answer another two questions: (1 – Feedback, FDB) In how far do you agree that the feature is a standard feature for mobile application development (1 = do not agree at all, 5 = fully agree)? (2 – Effort estimation, EES) How high would you estimate the effort for programming respectively conceptualizing the feature in man-days for each of the three scenarios best case, likely case and worst case?

We questioned 18 experts aged 20-49, of which 14 were male and four female. All of them worked for the IT service provider. The participants' work experience ranged from less than one year to five years, whereas 56% of them had an experience of one to two years. 39% of the participants had worked in less than five projects, 33% in six to ten projects and 27% in more than ten projects. The share of agile methods that had been used in these projects varied strongly: While 17% of all experts solely had worked with agile methods, 61% stated that they had used agile methods in less than half of their projects. Six of the participants were user experience designers and eight were programmers, while four worked in other fields. While in the first round of the survey all 18 participants took part, the number decreased to six in the second round.

4 Results

The first survey round focused on the frequency of implementation of the different features, the degree of customization and the localization of efforts. The results of the first two questions are shown in Table 1. The two features registration and login as well as chronological lists were rated as most often implemented, while messaging, sharing and navigation received the lowest values.

Table 1. Mean (sd) values for the Perceived Frequency (PFQ, 1 = never, 5 = everytime) and Degree of Customization (DOC, 1 = very little, 5 = very strongly) of the 16 app features.

Feature	PFQ	DOC
Registration and login	4.11 (0.66)	2.69 (1.26)
Chronological lists	4.11 (1.13)	3.40 (1.12)
Search	3.22 (0.68)	3.58 (1.31)
Push notifications	3.18 (0.56)	3.36 (0.96)
Calendar and time	3.06 (0.65)	2.69 (0.98)
Hardware access	3.06 (0.88)	3.00 (1.25)
Reservation and booking	2.78 (0.99)	4.27 (0.90)
File upload	2.72 (0.92)	3.50 (1.07)
Favorites	2.65 (0.87)	3.55 (0.82)
Map view	2.61 (0.99)	3.50 (1.41)
Localization	2.53 (1.01)	2.13 (0.81)
Scanning feature	2.38 (0.98)	2.29 (0.94)
Comment feature	2.28 (0.73)	3.20 (0.71)
Messaging	2.11 (0.87)	3.83 (1.23)
Sharing	2.00 (0.60)	2.33 (0.70)
Navigation	1.94 (0.85)	2.00 (0.75)

Regarding the degree of customization, reservation and booking is the feature that has the highest need of adaptation to customer requirements. Likewise, messaging, search and favorites are associated with a high degree of customization. Regarding sharing, localization and navigation, the rating was quite low.

As shown in Fig. 1, the localization of effort varied between conceptualization and programming. The horizontal marker position in this figure represents the mean value of the responses for each feature (1 = exclusively in conceptualization, 5 = exclusively in

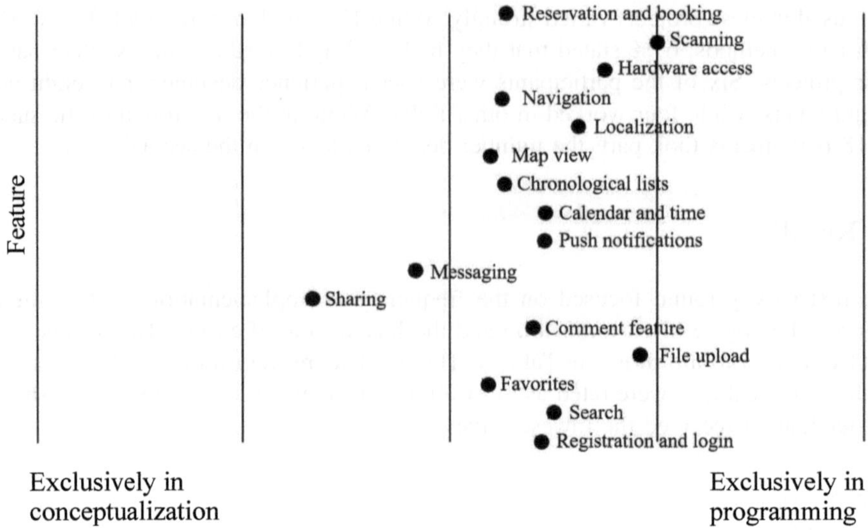

Fig. 1. Localization of effort for the app features.

programming). According to the participants, the most effort in app development is caused by programming, with scanning, file upload and hardware access receiving the highest values. Messaging and sharing are the only two features where the effort was mainly localized in conceptualization. In the second round of the survey, after collecting feedback for validating ten standard app features, effort estimations for these features were made, considering different efforts for conceptualization and programming.

In the second round, the participants were confronted with results from the first round. The two features registration and login (M = 4.83) and search (M = 4.17) were clearly confirmed as standard features for mobile apps by the participants. For all other features, the agreement ranged from M = 3.20 to M = 3.67, indicating slight confirmation, except for the reservation and booking feature, which received the lowest values (M = 2.50). The effort for programming was rated higher than the effort for conceptualization for nearly all features. Only for chronological lists was the conceptualization effort (M = 4.44 man-days) higher than the estimated effort for programming (M = 4.39 man-days). On an average, the rating for programming was 2.43 man-days higher than for conceptualizing.

Figure 2 shows the results of the effort estimation for implementing the standard features in a best case, likely case and worst case. Estimated values ranged from 2.75 to 27 man-days. The map view received the highest values, followed by the reservation and booking feature. In contrast, calendar and time as well as login were rated least time-consuming and also are the features with the smallest variability between best case and worst case estimation. Corresponding to that, the features that received the highest effort estimation values are the ones with the highest variability.

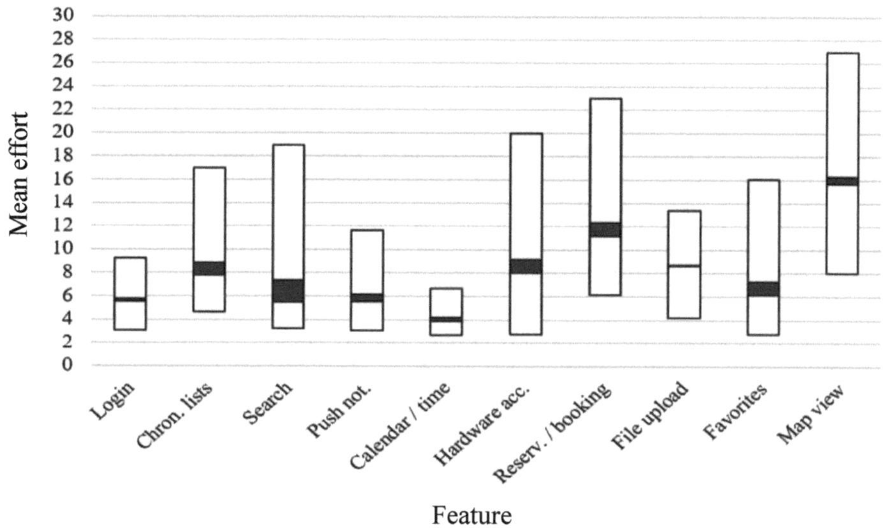

Fig. 2. Total expenditure for the standard features (*x-axis*), estimated in man-days (*y-axis*). The bars indicate the mean effort estimated in the best case scenario (*lower end of the bar*), the mean effort estimated in the worst case scenario (*upper end of the bar*) as well as the likely case estimations (*lower end of the shaded area*) and the weighed mean derived from the PERT-formula (*upper end of the shaded area*).

5 Discussion

We used a Delphi study approach for identifying and validating standard features for mobile apps in large enterprises using the example of the German IT service provider from the logistics and transportation sector. We used an expert panel based approach to identify frequently used app features. Localization of effort varied across features. In general, our approach was also capable to reveal effort estimates for all features. As mentioned before, we focused on the feasibility and could not provide data for a statistical verification of the results. Even when implemented in a large corporation, it is not very likely – due to the small sample sizes – that the available data will be sufficient to derive statistically verified effort and cost estimates. However, high levels of standard deviation or large ranges of effort estimates could be seen as an indicator to cluster for more homogenous project types (as frequency of feature use might depend on the type of project) and split features in sub features (to derive more precise estimates).

When looking at the results in more detail, in the first survey round, registration and login as well as chronological lists were stated to be the most often implemented features and therefore could be considered as standard features. In the second round of the survey, registration and login was clearly validated as a standard feature, while instead of chronological lists, the search feature received the second highest rating. Thus, we assume that all three are key features of mobile apps. As a key finding, our approach proved suitable, but nevertheless should comprise several survey rounds for validation, since the results showed that implementation frequency changes. The degree of customization was lowest for scanning, localization and navigation, what may be due to the fact that these features mainly require access to hardware components of the mobile device, such as cameras or sensors, or are realized by means of standardized user interface elements and therefore are characterized by less varying parameters. The results indicate that only few features need to be adapted to customer requirements particularly highly or particularly little. In fact, almost all features received medium ratings regarding their degree of customization. For getting to the bottom of this aspect, further research on different characteristics, reasons and implications of customization is needed. In general, the main effort for app development is localized in programming. This is also supported by the fact that, based on our results, programming requires more effort than conceptualizing for almost all app features. Nevertheless, the values differ strongly between both perspectives and for each feature. Given these observations, another key finding of our study is that differentiating between several perspectives and roles is essential for effort estimation in mobile app development and needs to be taken into account. Furthermore, several features showed a high variance regarding the effort in a best case, likely case and worst case scenario. Since the big variance may be caused by either a high complexity of these features or the summarizing of similar features during data set preparation, further research is needed that concentrates especially on these features.

As a next step in our research, we plan to conduct studies that compare effort estimates derived from our Delphi-like approach with estimates for the features estimated based on traditional software cost estimation methods. A comparison of these values could give some indication of the expected level of accuracy for our results. We are also planning to generate total effort estimates based on the feature estimates and compare them with available information for completed app projects.

6 Managerial Implications

In order to utilize the proposed effort estimation method, we are working towards a prototypical calculation tool that builds on our findings. The tool shall be designed as a web application or desktop software and is meant to provide an overall estimate as well as some more detailed figures on the effort drivers (e.g., design, programming). It also needs to abstract from the underlying calculation logic to provide an easy use and intuitive estimation process aligned to today's agile and user-centered app development methods. This also ensures that even non-experts in the field of software cost estimation are able to conduct an estimation. Furthermore, the calculation base needs to be adjustable to the specific conditions of each project. This includes the respective effort values of each app feature and the hourly wages of the involved employees.

Taking into account the challenges of mobile app development and the resulting issues for effort and cost estimation in this field, the tool needs to allow estimation based on incomplete information on the one hand and recurrent estimation throughout the project on the other hand. Therefore, the core function of the tool is to enable the user to make continuous updates or refinements. Accordingly, the tool must allow a modification of the initial estimates at any time. To revise the overall estimate, the user may need to add app features that were not part of the initial estimate, remove components that are no longer part of the project or to reassess the scaling factors and cost drivers. This happens at three stages: (1) Before the project starts, (2) during the project, and (3) after the project is finished. In the first stage, a new project is created in the tool. This could be done by either directly selecting features from a repository or by analyzing use case description and recommending features that have been used to support this use cases in completed projects. On this basis, the tool could automatically calculate a first effort estimate. In the second stage, during the project, the tool could also be used to foster collaboration across projects by providing organizational data (e.g., the information on a developer within the company who worked on a similar feature before). Furthermore, app features can be removed or added according to evolving requirements throughout the project and hence, the total expenditure is automatically adapted by the tool. In the third stage, effort values can be saved or corrected, corresponding to the actual costs and efforts spent after finishing the project, in order to improve the reliability and accuracy of the estimates. In addition, reasons for deviations from an effort estimate could be reported when this information is available.

7 Conclusion

We proposed an iterative and experience based approach to estimate the effort for app development in large enterprises along agile and user-centered projects in the absence of detailed requirements specifications. The approach is based on app features and was evaluated for basic feasibility in a cooperating German IT service provider using an adaptation of the Delphi method. The results showed that our approach was able to identify recurring app features as well as effort estimates for these features. The study revealed that different perspectives and roles involved in the development process have to be taken into account, since the estimates differ between UX designers and programmers. We also proposed some basic considerations for a concept of a tool that utilizes these findings and could be implemented in practice. The core function of the tool is to enable the user to continuously update and refine the estimated effort and cost values in mobile app development, accompanying the project. Limitations of our work are the evaluation within one single company and the low number of participants. Further research is required to analyze the accuracy of the derived estimates within the company and compared to more traditional software effort estimate methods. Moreover, an evaluation across companies and the effect on accuracy of a continuous data collection process (instead of a single Delphi study) has to be investigated.

Acknowledgements. This work was funded by the German Federal Ministry of Education and Research, grant no. 03FH032PX5; the PROFRAME project at RheinMain University of Applied Sciences. All responsibility for the content of this paper lies with the authors.

References

1. Norman, D.A., Draper, S.W. (eds.): User Centered System Design. New Perspectives on Human-Computer Interaction. Erlbaum, Hillsdale (1986)
2. Corral, L., Sillitti, A., Succi, G.: Software development processes for mobile systems: is agile really taking over the business? In: 1st International Workshop on the Engineering of Mobile-Enabled Systems (MOBS), pp. 19–24 (2013)

3. Boehm, B.W.: Software Cost Estimation with Cocomo II. Prentice Hall, Upper Saddle River (2000)
4. Ferrucci, F., Gravino, C., Salza, P., Sarro, F.: Investigating functional and code size measures for mobile applications. In: 41st Euromicro Conference on Software Engineering and Advanced Applications, pp. 365–368 (2015)
5. Heemstra, F.J.: Software cost estimation. Inf. Softw. Technol. **34**(10), 627–639 (1992)
6. Nitze, A., Schmietendorf, A., Dumke, R.: An analogy-based effort estimation approach for mobile application development projects. In: Joint Conference of the International Workshop on Software Measurement and the International Conference on Software Process and Product Measurement, pp. 99–103 (2014)
7. Abdullah, N.A.S., Rusli, N.I.A.: Reviews on functional size measurement in mobile application and UML model. In: Proceedings of the 5th International Conference on Computing and Informatics, pp. 353–358 (2015)
8. Jurisch, M., Igler, B., Böhm, S.: PROFRAME: a prototyping framework for mobile enterprise applications. In: CENTRIC 2016, The Ninth International Conference on Advances in Human oriented and Personalized Mechanisms, Technologies, and Services, pp. 7–11 (2016)
9. Jorgensen, M., Shepperd, M.: A systematic review of software development cost estimation studies. IEEE Trans. Softw. Eng. **33**(1), 33–53 (2007)
10. Schmietendorf, A.: Kosten- und Aufwandsschätzung bei der Entwicklung von Smartphone-Apps, eine Bestandsaufnahme. In: MetriKon 2012 - Praxis der Software-Messung, pp. 141–148. Shaker, Aachen (2012)
11. Nitze, A.: Measuring mobile application size using COSMIC FP. In: MetriKon 2013 - Praxis der Software-Messung, pp. 101–114 (2013)
12. Jošt, G., Huber, J., Heričko, M.: Using object oriented software metrics for mobile application development. In: 2nd Workshop of Software Quality Analysis, Monitoring, Improvement, and Applications, pp. 17–27 (2013)
13. De Souza, L.S., de Aquino Jr., G.S.: Estimating the effort of mobile application development. In: 2nd International Conference on Computational Science and Engineering, pp. 45–63 (2014)
14. Francese, R., Gravino, C., Risi, M., Scanniello, G., Tortora, G.: On the use of requirements measures to predict software project and product measures in the context of android mobile apps: a preliminary study. In: 41st Euromicro Conference on Software Engineering and Advanced Applications, pp. 357–364 (2015)
15. otreva. https://www.otreva.com/calculator/
16. Kinvey, Inc.: http://calculator.kinvey.com/mbaas-savings-calculator
17. New Signature. https://newsignature.com/mobile-app-calculator/
18. Apprasoft. https://apprasoft.com/app-estimator/
19. Boehm, B.W.: Software Engineering Economics. Prentice Hall, Englewood Cliffs (1981)
20. Boehm, B.W., Clark, B., Horowitz, E., Westland, C., Madachy, R., Selby, R.: Cost models for future software life cycle processes. COCOMO 2.0. Ann. Softw. Eng. **1**(1), 57–94 (1995)
21. Amberg, M., Bodendorf, F., Möslein, K.M.: Wertschöpfungsorientierte Wirtschaftsinformatik. Springer, Heidelberg (2011)
22. Broy, M., Kuhrmann, M.: Projektorganisation und Management im Software Engineering. Springer, Berlin (2013)

Assessing the Effectiveness of Emoticon-Like Scripting in Computer Programming

Angelos Barmpoutis(✉), Kim Huynh, Peter Ariet, and Nicholas Saunders

Digital Worlds Institute, University of Florida, Gainesville, FL 32611, USA
angelos@digitalworlds.ufl.edu,
{huynhtina123, pjpariet, password1}@ufl.edu

Abstract. In this paper a new method is proposed for learning computer programming. This method utilizes a set of human-readable graphemes and tokens that interactively replace the grammatical tokens of programming languages, using a concept similar to emoticons in social media. The theoretical framework of the proposed method is discussed in detail and two implementations are presented for the programming language ECMAScript (JavaScript). The results from user testing with undergraduate students show that the proposed technique improves the student's learning outcomes in terms of syntax recall and logic comprehension, in comparison to traditional source code editors.

Keywords: Computer education · Programming languages · Human factors · Interaction design · Emoticons · Graphemes

1 Introduction

In many ways, learning to program can be challenging for beginners of all ages. This is often attributed to poor self-efficacy, limited prior experience with computers, or inability to relate personal experiences to abstract programming concepts [1]. In turn, these challenges often result in beginners to regard programming as "boring and difficult" [2]. Since the early 1960's, there have been a number of tools developed to address these issues and overall focused on making programming accessible for everyone. With the wide array of modern Computer Science (CS) educational tools available, identifying the benefits and challenges associated will be necessary to gain insight into the varying effectiveness of each tool [3–5]. Currently, the two most notable types of interactive tools for computer science education are Tangible User Interfaces (TUI) and Graphical User Interfaces (GUI).

TUI systems are in the form of physical objects and environments, which can represent or be augmented by digital information [6]. The biggest advantage of TUI systems is that they build directly on existing knowledge and experience from the real world [4]. One example of a TUI system is Robo-Blocks, which enable users to program a moving robot by connecting blocks that contain embedded technology, where each block represents a certain command [7]. Studies have shown that the use of Robo-Blocks show considerable potential in helping children learn how to debug, and further develop problem solving skills [5, 7]. Another example is Tern, which is a set of

T. Ahram and W. Karwowski (eds.), *Advances in Human Factors, Software, and Systems Engineering*, Advances in Intelligent Systems and Computing 598,
DOI 10.1007/978-3-319-60011-6_7

wooden blocks, where each block represents a basic command that can be connected with others in order to create a program [8]. Through minimizing the amount of resources used in creating Tern, it provides a practical way for children to have real-time interactions with CS concepts [9]. Studies have shown that children had noticeably high hands-on interaction when using Tern, which indicates that it can encourage children to take more active roles in learning [8, 9]. However the major disadvantage of TUIs is their disconnect from programming languages and thus cannot be used beyond an early stage of learning.

GUI programming tools employ virtual graphic components as means for interaction with a computer, which have attracted in general a wider demographic variety of users than TUIs, since GUIs have greater flexibility in content and have the capability to improve user's understanding at a university level. One example of a GUI system is Alice, which is a programming environment that enables novices to develop 3D environments using drag and drop scripts [10, 11]. Studies on the use of Alice in introductory CS courses have shown that the average student scores were consistently higher, when compared to those in courses using Java [10]. Scratch is another example of a GUI system, which allows users to program "interactive stories, games, and animations" through an online platform [12]. User studies indicate that middle school and college students found Scratch to be a reliable way to introduce beginners to the basics of programming, because it removed the complexity of syntax [12, 13]. However, in the same studies, a number of students did not favor Scratch because it underestimates the detail and complexity of more comprehensive programming languages. While GUIs can provide a reliable starting point, the students should transition into a more advanced programming language [12], which is the main limitation of this method.

In this paper, we propose a new educational framework that overcomes the problems of the aforementioned approaches by adding a human-readable layer on the top of existing programming languages. The proposed method is based on the use of emoticon-like typing that has become popular with social networks. Emoticons are visual representations that have one to one relationship with a corresponding combination of characters such as ":)". These can be perceived as visual interpretations of the corresponding characters that provide instant feedback to the user regarding the meaning associated with the typed code. The proposed framework utilizes a set of meaningful visual replacements of each grammatical token in a given programming language that appear instantly when complete valid tokens are typed.

The proposed method, dubbed Brain-Activating Replacement method (BAR), is based on the following three hypotheses: (a) the immediate feedback given to the programmer can result in improved learning outcomes as it stimulates the brain to build one-to-one connections, (b) the unique correspondence of each visual replacement, with a valid programming token re-enforces the learning of the syntax in an intuitive trial-and-error framework, (c) the use of visual replacements remove visually the grammatical and syntactical details of a programming language and reveal to the users the logic of the program in the form of a pseudo code.

The developed framework was evaluated in a pilot study using 35 undergraduate students, and the results conclusively demonstrate the merits of the proposed method.

2 Methods

The smallest units in any writing system are known as graphemes [14]. Graphemes are not only the characters in a given alphabet but also the accents, punctuation marks, and other symbols that may be used in the corresponding writing system. Similarly, in any programming language a set of graphemes is used, which usually includes the graphemes of the Latin alphabet as well as other logical, mathematical, and structural symbols required for the needs of a particular programming language.

One or more graphemes can form morphemes, which are the smallest grammatical units in a particular language. One or more morphemes can function as components of a word, which along with various self-standing graphemes and morphemes are the smallest self-standing units in a language, also known as tokens. Morphemes contribute a particular meaning to a token when used as suffix, prefix, or as an intermediate compound of the token. For example, in computer programming the morphemes "+" and "=" can form the token "+=", or they can function as independent tokens by themselves. In this example, the token "+" denotes addition, the token "=" denotes assignment, and the token "+=" denotes "be increased by", which contains addition and assignment.

This hierarchical structure of written languages continues in higher levels in order to compose more complex constructions as ordered compositions of lower level units, such as sentences, paragraphs, sections, chapters, books, and book series. Similarly, in computer programming one or more tokens when used in the proper order can form a command. Subsequently, one or more commands can form a self-contained program, such as a function. Finally, one or more functions can compose larger structures such as classes, or an executable computer program.

Let Ω_0 be the set of graphemes in a written language. A morpheme can be expressed as an ordered set of graphemes as follows:

$$\mu = \{\gamma_1, \gamma_2, \gamma_3, \cdots : \gamma_i \in \Omega_0\} \tag{1}$$

If Ω_1 is the set of all morphemes in a written language, a token can be expressed similarly as:

$$t = \{\mu_1, \mu_2, \mu_3, \cdots : \mu_i \in \Omega_1\} \tag{2}$$

where $t \in \Omega_2$, which is the set of all tokens in a given language. Equations 1 and 2 can be generalized in order to account for higher grammatical levels as follows:

$$\{e_1, e_2, e_3, \cdots : e_i \in \Omega_{k-1}\} \in \Omega_k \quad \forall k > 0 \tag{3}$$

where e_i is an element of the previous hierarchical set. For example, in the case of k = 1, e_i represents a grapheme, in the case of k = 2, e_i represents a morpheme, in the case of k = 3, e_i represents a token, etc.

2.1 Human-Readability of Programming Languages

One key difference between written natural languages and programming languages is that the former are constructed as representations of spoken languages, and for this reason, the formation of graphemes and morphemes imitates the corresponding phonemes of the spoken language. This correspondence between graphemes and phonemes makes a written text easy to be read and comprehended by speakers of the corresponding spoken language. However, such correspondence does not exist in programming languages, since their graphemes were not derived by a phonetic system but were specifically defined in order to facilitate the communication with computers. In that sense, the source code of a program is meant to be natively read by a computer rather than humans, who can in turn decipher the corresponding programming context but can only communicate it after having interpreted it to a human spoken language.

Therefore, the primary role of graphemes and tokens in a programming language is to serve as input to computers rather than as output to humans. This leaves a significant gap in the process of writing/reading a computer program, which is more evident in the case of beginner programmers who often try to read and comprehend a given text written in a computer programming language. For example the text "a += 2;" in many programming languages means "increase the value of a by 2". By comparing the original phrase with the translated one, it is evident that the latter is easier to read and understand especially in the case of beginner programmers.

In this paper, a new text-editing process is proposed as a solution that bridges the aforementioned gap in computer languages. The proposed solution does not intend to form a new programming language but enhance the readability of existing ones by extending the traditional human-computer interaction of text editors.

Let us consider the following written sample: "not:(or:|!be:)" and its equivalent in another written language with different graphemes: "not☹or😐!be☺. Obviously, the latter is easier to read, but the former is easier to write in the form of a typed text in a computer device. This example shows that there exist written languages that are primarily meant to be written (possibly to serve as an input to a computer system), and others that are primarily meant to be read. Furthermore, there exists a mapping that maps elements of the former language to elements of the latter:

$$\mathrm{M}_k : (\Omega_k, \Omega_{k+1}) \rightarrow \Psi_k \tag{4}$$

where Ω_k denotes the set of elements in the kth hierarchical level of one language, and Ψ_k denotes the set of elements in the same hierarchical level of another language. According to the previous example, M_k (":)", C) = "☺", where k is the token-level and C denotes the context set C = {"*not*", ":(", "*or*", ":|", "!", "*be*", ":)"}, which is an element of the next hierarchical level $k + 1$.

The role of the context Ω_{k+1} in Eq. 4 is to enable us to define context-depended mappings for a particular element Ω_k. For example, the token "=" in ECMAScript (JavaScript) programming language can be mapped to "be:", unless it is followed by the token "[", in which case it can be mapped to "be the following array:". Finally, Eq. 4 can be further generalized in order to consider the context in different hierarchical levels as follows:

$$\mathrm{M}_k : (\Omega_k, \Omega_{k+1}, \Omega_{k+1}, \cdots) \rightarrow \Psi_k \quad (5)$$

Equations 4 and 5 can be used in order to provide readability to computer programming languages without altering in any way the programming languages themselves. A set of mappings M_k and a target language Ψ must be defined according to Eqs. 4 and 5. Different mappings may be defined in different hierarchical levels. For instance, the previous example in ECMAScript defined the following token-level mapping: {“a”, “=”, “2”} → {“a”, “be:”, “2”}. An additional command-level mapping can be defined in order to map {“a = 2”} → {“Set a to 2”}. The corresponding mapping can be used as soon as the required input elements of the original language Ω are typed. For example, during the composition of a token, a grapheme-level mapping can be used. Once a complete token is composed, a token-level mapping can be employed in order to render the token. Similarly, a command-level mapping can be used as soon as a complete command is composed.

Obviously, the result of mappings can be modified or cancelled when the input that activated these mappings is changed. For example, when a previously typed command is being edited, the command-level mapping is cancelled, and a token-level mapping is used to render the tokens of this command. Similarly, when the user edits a previously typed token, the token-level mapping is cancelled, and a grapheme-level mapping can be activated to render this token.

2.2 Properties

The theoretical framework presented in the previous sections for interactively replacing the elements of an input written language Ω with elements of a target language Ψ has the following properties:

Preservation of Cardinality. The mapping from Ω to Ψ should preserve the number of elements in each level, i.e. N tokens in Ω should be mapped to N tokens in Ψ, M commands in Ω should be mapped to M commands in Ψ, etc.

Interpretative Replacements. The elements of the target language can employ graphemes, symbols, or other textual representations that interpret the corresponding elements of the original language. The replacements can be properly chosen based on the age, proficiency in Ω, or personal preferences.

Interactive Validation. The instant replacement of elements of Ω provides continuous feedback that validates the user's input and offers a trial-and-error interface for text composition.

Authentic Reproduction. A text in Ψ cannot be reproduced unless the original text is re-typed in Ω, which requires that the user is capable of composing the original text in Ω. Therefore, it is an authentic reproduction since the user cannot blindly copy the text in Ψ in order to achieve the desired result.

Syntax Discovery. The combination of authentic reproduction and interactive validation facilitates the discovery of syntactical phenomena and is hypothesized to re-enforce learning by stimulating the brain to build connections between the programming language (Ω) and its interpretation (Ψ).

Identity Mapping. Traditional text editors can be considered special cases of the proposed framework in which the identity mapping is used between Ω and Ψ. Source code editors that utilize color-coding are also special cases, in which the graphemes in Ψ differ from those in Ω only in their color properties and not in their structure.

The aforementioned properties extend the traditional text editing process by introducing new interactive features that do not currently exist in source code editors, but the general audience has established familiarity with them through emoticon scripting in social media and text messages.

3 Implementation

The proposed method, dubbed Brain-Activating Replacement (BAR) method, was implemented for ECMAScript (JavaScript) in two different target languages. The first implementation was designed for beginner programmers and employs iconic graphemes and visual metaphors such as nametags for the names of variables and pipes for representing functions. The second implementation was designed for more mature users and employs more discrete iconic graphemes with primarily text-based token replacements. For each implementation, a set of 88 BAR-tokens was designed. Table 1 shows a sample list of the tokens from the target language designed for beginner programmers.

As shown in Table 1, all data values are visualized inside rectangular boxes that resemble fields in an electronic form. Comparison and logical operators as well as structural markers are replaced by textual interpretations, such as “begin”, “is not”, etc. An example of JavaScript source code visualized in the two implemented target languages is shown in Fig. 1.

By comparing the output of the two implementations as shown in Fig. 1, it is evident that the implementation designed for beginners is more colorful and iconic, compared to the one designed for mature audiences, which is primarily text-based and

Table 1. A sample list of the BAR-tokens created for beginner programmers. The table shows 19 out of the 88 types of tokens implemented for ECMAScript (JavaScript).

variable	*name*	variable being modified	name
number	123.45	string	'a string'
true, false	✓ ✗	null, undefined	???
;	↵	method()	method
=	**be:**	==, !=, ===	**is equal to, is not, strictly is**
var	**Let**	{ }	**begin, end**

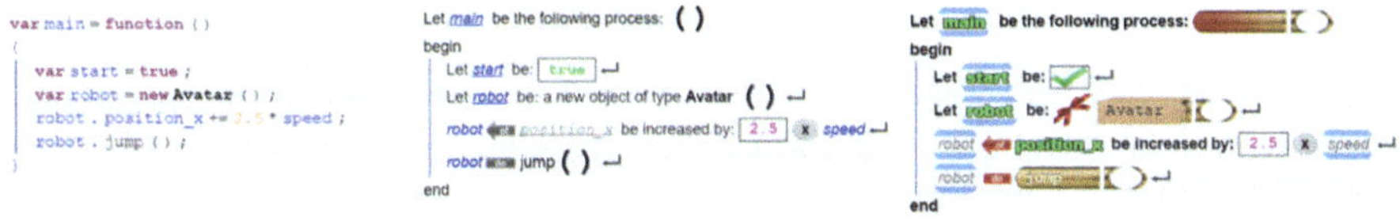

Fig. 1. Side-by-side comparison of the two implemented target languages for beginners (*right*) and for more mature users (*middle*). The corresponding JavaScript source code is on the left.

can be read almost as a continuous text: “Let main be the following process: Begin. Let start be true. Let robot be a new object of the type Avatar. Position_x of robot be increased by 2.5 x speed. Robot do jump. End.” One interesting observation is that the tokens “=” and “.” were interpreted differently based on their context as defined in Eq. 5. For example “.” was interpreted as “do” when followed by a method and as “of” when followed by a property.

The two implementations (Fig. 1 middle and right) and the traditional implementation (Fig. 1 left) were developed in JavaScript using the open-source library VisiNeat, which is licensed under BSD 2-clause by the University of Florida. The developed BAR-enabled text editor is available in the VN JavaScript Studio and was used in a series of experiments discussed in the next section.

4 Experimental Results

The proposed BAR method was tested during the academic semester of Spring 2017, using students volunteers from the on-line and on-campus undergraduate program of Digital Arts and Sciences at the Digital Worlds Institute at the University of Florida, with partial support from the University of Florida On-Line Institute and the grant PRDSP024 from the University of Florida Provost’s office.

In total 35 students (22 female) used the developed text editor to complete 6 programming assignments for a period of 7 weeks. During this period, the proposed method was adopted as the primary method of instruction; hence, all the content delivered to the students as part of lectures or additional material was in the form of BAR-tokens. It should be noted that the students were allowed to choose between the two developed BAR-enabled text editors (Fig. 1 middle and right) and a regular source code editor (Fig. 1 left), or transition between the editors as they wished.

The students in this program focus on the theory and practice of interactive digital media, and, as part of their curriculum, learn programming fundamentals. Although the majority of the students have limited prior experience in programming, others have taken prior programming classes or practiced programming on their own. It was anticipated that the proposed framework would have different effect on the students based on their programming level, as it is demonstrated later in this section.

4.1 Manual Observation of Student Coding Patterns

During the course of this experiment, the keystrokes performed by the participating students within the developed text editors were recorded in order to manually inspect the coding patterns of the students with or without the proposed framework. After systematic inspection of the recorded sequences of keystrokes, the following scenarios were noted.

The students who typed their assignments in a BAR-enabled text editor were able to identify and correct syntax errors as well as discover unknown syntax rules on their own before the compilation of their code. A common mistake was the use of spaces within the name of a variable as shown in Fig. 2. The corresponding BAR tokens provided instant visual feedback that assisted the students to identify and correct their mistakes.

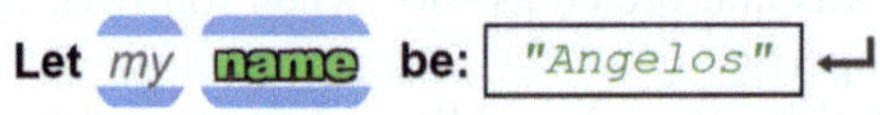

Fig. 2. This figure shows an example of erroneous use of spaces within the name of a variable. The corresponding BAR-tokens provide instant visual feedback in order to self-correct this error. In this case, 2 nametags are shown and only the second one is affected by the assignment.

On the other hand, students who made similar mistakes in a traditional text editor were not able to identify their mistake prior to compilation. What is more, there were several instances where the students could not understand their mistake even after receiving an "unexpected identifier" error message from the compiler.

Another mistake that was observed several times is the incorrect use of the equal sign in an attempt to test equality instead of the double equal sign "==", which is the appropriate token to be used in this case. The problem in this scenario is that this is a logical rather than a syntax error; hence, the compiler does not provide any error message to help the programmers identify their mistake. However, in the case of BAR tokens, there is a clear difference between the correct and incorrect case of testing equality. More specifically, in the correct test of equality the target code reads "if(score is equal to 100)", which corresponds to the source code "if(score == 100)". On the other hand, when the source code is "if(score = 100)", the target code reads "if(score be: 100)", which reveals the incorrect logic. It has been recorded several times in the collected data that the students who used the BAR-enabled editor were able to identify the difference between these two statements and correct their logic without further assistance.

4.2 Quantitative Analysis

Several types of data were recorded during this experiment, including the type of text editor used in each keystroke and the error messages generated by the system (syntax errors and run-time errors), in addition to the keystroke sequences produced by each student.

Figure 3 shows the transitions of the students between the provided text editors during the data collection period. As expected, a gradual transition from the

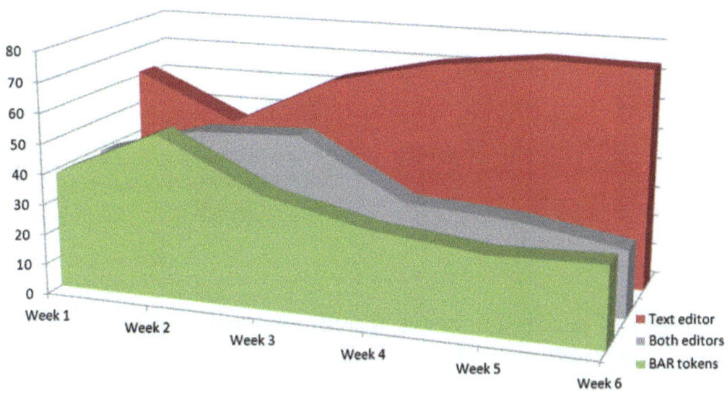

Fig. 3. The percentage of the students who used the proposed editor and/or traditional code editor during this experiment. The plot shows the transitions of the students between editors.

BAR-enabled editor to the traditional editor was observed as the competency of the students in programming increases, although a small transition back to the BAR-enabled editor was also observed towards the end of the data collection period. It should be noted that a significant percentage of the students completed their programming assignments using both editors.

The syntax and run-time error messages generated by each student are shown in Fig. 4. The data were normalized by the number of keystrokes per student in each type of editor. Although the absolute value of the reported numbers depend on the particular level of difficulty of each week's programming assignment, we can compare the data across categories. By observing the data in Fig. 4, it should be noted that the number of run-time errors generated by the students who used the BAR-enabled editor is less than the corresponding number of errors from the traditional text editor, and this is true throughout the entire data collection period. A similar pattern was observed for syntax errors with the exception of weeks 2 and 4. This observation may indicate that the effect of the proposed method on logic understanding is stronger than the effect on syntax recall, since the majority of the run-time errors are typically associated with logic errors.

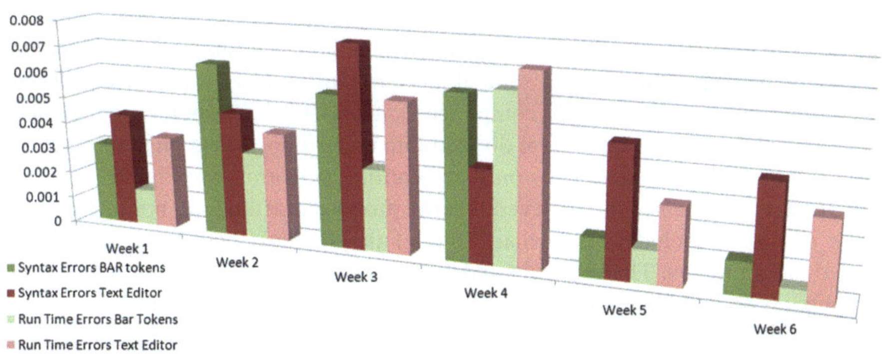

Fig. 4. The recorded number of syntax and run-time errors per keystroke per person during this experiment. The results are reported separately for the proposed and traditional text editors.

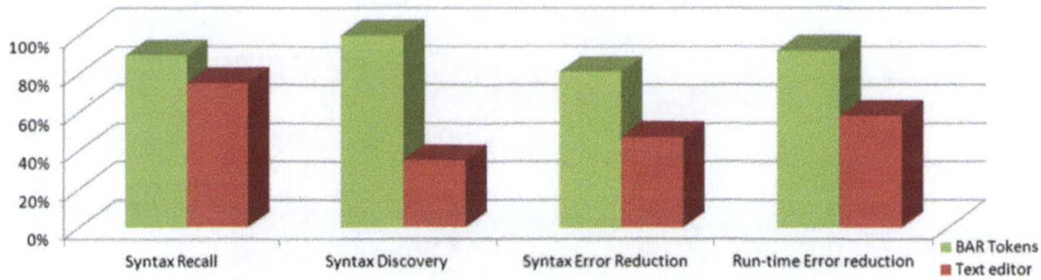

Fig. 5. The percentages of successful syntax recall, new syntax discovery, and error reduction that correspond to the proposed method versus traditional text editors.

At the end of the data collection period, the students were asked to recall particular syntax rules as well as try to comprehend unfamiliar syntactical structures. Figure 5 shows that 90% of the students who used BAR-tokens were able to successfully recall the syntax compared to 75% for the case of a traditional text editor. This result suggests that the students who see the typed source code are not able to remember the syntax as effectively as the students who do not see the typed code.

Furthermore, only 30% of the students who read an unfamiliar sample of code in a conventional text editor were able to comprehend it. The same task in the BAR-enabled editor was reduced to the study of a text in plain Engish, which was comprehended by all students.

Finally, Fig. 5 reports that the overall reduction of errors using the BAR method was larger than the corresponding errors in a traditional text editor. If the reduction of errors is assumed to be correlated with the learning outcomes, then the results may suggest that higher learning outcomes can be achieved using the proposed method.

4.3 Perceived Usefulness and Perceived Ease of Use

Finally, at the end of the data collection period, the proposed method was evaluated using the technology adaptation model (TAM) [15]. The original questionnaires of the TAM model were extended in order to capture the perception of the students regarding the effect of the proposed technique on syntax recall and logic comprehension.

In order to evaluate the effect of the proposed method as a function of the student's competency in programming, each student's programming level was assessed through programming questions. Principal component analysis of the responses on the TAM survey showed that the students were divided into two groups based on their perception regarding the effect of BAR-tokens on their learning performance (Fig. 6).

The group of students who believed that the proposed method did not improve their learning outcomes includes more than 70% of the students who were identified as advanced programmers. Therefore, the independent analysis of the two groups can give us insights on how students of different programming levels perceived the usefulness and ease of use of the proposed method.

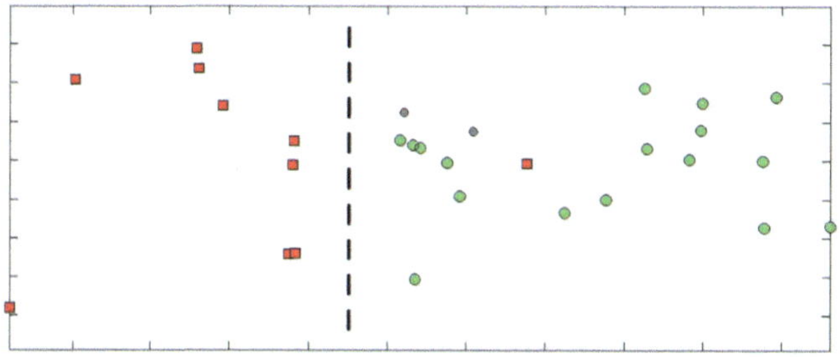

Fig. 6. The distribution of the students based on their responses on the TAM questionnaires as shown on the dominant eigen-plane of the response data. The students can be linearly separated based on their perception regarding the improvement of their learning performance, depicted in green and red for positive and negative responses respectively.

Figure 7 shows that the majority of students on both levels agreed that the proposed method challenges them more to learn the syntax. This result aligns with the "authentic reproduction" property of our theoretical model as discussed in Sect. 2.2.

Furthermore, the students in group A, who are predominantly beginners in programming, agreed on the majority of the questions, as they found that they could

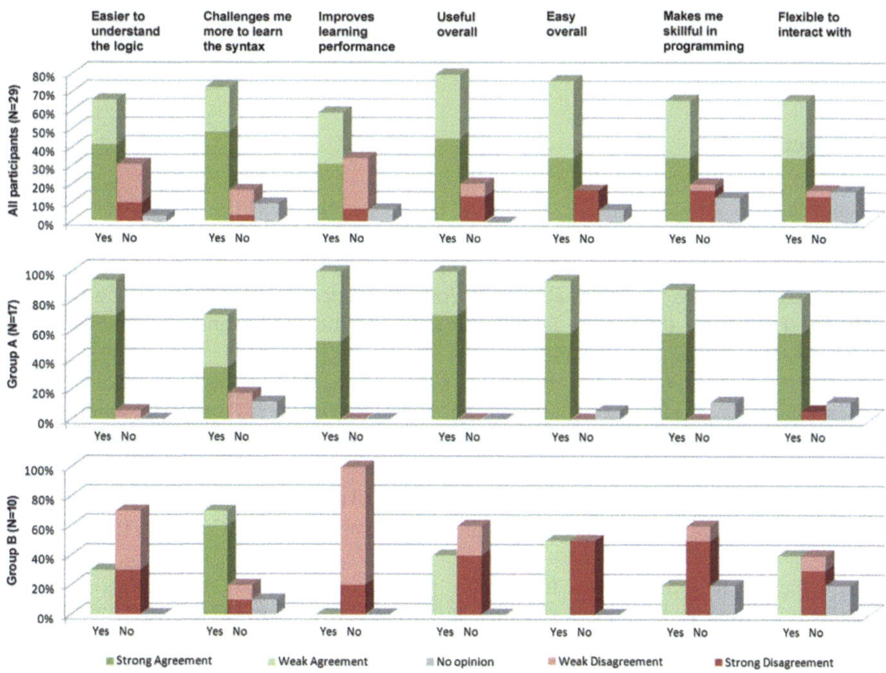

Fig. 7. The students' responses on the key questions of the extended TAM survey [15]. The results are also shown separately based on the students' perception regarding the improvement of their learning performance (*middle row: positive response, bottom row: negative response*).

understand the programming logic easier with the proposed method. Furthermore, they believed that the proposed method improved their learning performance and can make them skillful in programming. They also found that the proposed method is useful, easy, and flexible to interact with.

On the other hand, the students in group B, who are predominantly more experienced programmers, appeared to be divided on several questions regarding the flexibility and ease of use of the proposed method. As it was expected, the effect of the proposed method on the more advanced students was limited, which is reflected on the students' perception on the ease of use and usefulness of the proposed method.

5 Conclusions

This paper presented a novel method for learning computer programming, dubbed Brain-Activating Replacement method, which is based on the hypothesis that the interactive replacement of syntactical tokens in programming languages with human-readable tokens, facilitates the building of stronger connections between the source code and its logical meaning. Two implementations of the proposed method were presented and tested in a pilot study. The results suggested that the proposed framework increases the learning outcome of the students. The observed benefit is stronger in the case of beginner programmers, whose performance has improved in terms of syntax recall and logic comprehension, compared to the performance achieved using traditional text editors for source code editing.

References

1. Lahtinen, E., Ala-Mutka, K., Järvinen, H.M.: A study of the difficulties of novice programmers. ACM SIGSCE Bull. **37**(3), 14–18 (2005)
2. Jenkins, T.: On the difficulty of learning to program. In: Proceedings of the 3rd Conference of the LTSN Centre for Information and Computer Sciences, vol. 4, pp. 53–58 (2002)
3. Busch, T.: Gender differences in self-efficacy and attitudes towards computers. J. Educ. Comput. Res. **12**(2), 147–158 (1995)
4. Jacob, R.J.K., Girouard, A., Hirshfield, L.M., Horn, M.S., Shaer, O., Solovey, E.T., Zigelbaum, J.: Reality-based interaction: a framework for PostWIMP interfaces. In: Proceedings of the SIGCHI Conference on Human Factors in Computing Systems, pp. 201–210 (2008)
5. Sapounidis, T., Demetriadis, S., Stamelos, I.: Evaluating children performance with graphical and tangible robot programming tools. Pers. Ubiquit. Comput. **19**(1), 225–237 (2015)
6. McNerney, T.S.: From turtles to tangible programming bricks: explorations in physical language design. Pers. Ubiquit. Comput. **8**(5), 326–337 (2004)
7. Sipitakiat, A., Nusen, N.: Robo-blocks: designing debugging abilities in a tangible programming system for early primary school children. In: Proceedings of the 11th International Conference on Interaction Design and Children, pp. 98–105 (2012)

8. Horn, M.S., Solovey, E.T., Crouser, R.J., Jacob, R.J.: Comparing the use of tangible and graphical programming languages for informal science education. In: Proceedings of the SIGCHI Conference on Human Factors in Computing Systems, pp. 975–984 (2009)
9. Horn, M.S., Jacob, R.J.: Tangible programming in the classroom with tern. In: CHI 2007 Extended Abstracts on Human Factors in Computing Systems, pp. 1965–1970 (2007)
10. Moskal, B., Lurie, D., Cooper, S.: Evaluating the effectiveness of a new instructional approach. ACM SIGCSE Bull. **36**(1), 75–79 (2004)
11. Sykes, E.R.: Determining the effectiveness of the 3D Alice programming environment at the computer science I level. J. Educ. Comput. Res. **36**(2), 223–244 (2007)
12. Malan, D.J., Leitner, H.H.: Scratch for budding computer scientists. ACM SIGCSE Bull. **39** (1), 223–227 (2007)
13. Meerbaum-Salant, O., Armoni, M., Ben-Ari, M.: Learning computer science concepts with scratch. Comput. Sci. Educ. **23**(3), 239–264 (2013)
14. Coulmas, F.: The Blackwell Encyclopedia of Writing Systems, p. 174. Blackwell, Oxford (1996)
15. Davis, D.F.: Perceived usefulness, perceived ease of use, and user acceptance of information technology. MIS Q. **13**(3), 319–340 (1989)

Adaptive Interface Combined to Interface Patterns in Human Factors

Natalie Mie Takahashi(✉) and Plinio Thomaz Aquino Junior

Fundação Educacional Inaciana Pe. Sabóia de Medeiros,
Centro Universitário FEI, São Bernardo do Campo, São Paulo, Brazil
{natmt,plinio.aquino}@fei.edu.br

Abstract. Adaptive interface is an artifact with the objective to learn the user interactions and improve the system experience, and the adaptive system is a system that can adapt itself according with the user's goals and preferences. Patterns can be defined as solution to recurring problems in a specific context. The interface pattern are patterns that help to organize the interface, such as structure and behavior, in order to generate more understandable and usable interface. The Interface Patterns in Personas Layers (IPPL) are interface patterns that contain a structure organized in layers that are built and classified according with personas. The goal of this paper is to present a structure of interface patterns based on IPPL that allows future projects of adaptive interface and human-robot interaction to consider the automatic changes of the iterative process based on a set of patterns.

Keywords: Human-Computer Interaction · Human-robot interaction · Adaptive interface · Interface pattern · Human factors

1 Introduction

The Human-Computer Interaction (HCI) is a multidisciplinary field that studies the interaction of humans and computers, considering the human factors implications and the computational characteristics, where its main goal is to make systems more usable and accessible to people [1].

For this work, it is important to define the concepts of adaptive, adaptable and personalization in Human Factors. The adaptive is a property of the ability to change according with the conditions. From the system perspective, it is the automatic interpretation of the users' needs interpreted by the system. In this context, the term adaptable has the same meaning. Personalization is the automatic feedback of the system according with user's preferences [2]. Adaptive interface is an artifact with the objective to learn the user interactions to build user models in order to improve the system experience by mesmerizing the user's actions and improving the system interactions [3]. The adaptive system is a system that can adapt itself according with the user's goals and preferences. These properties are stored in a user model. The user model holds the information about the user's preferences to provide information that might be relevant for, distinguishing one from another [2]. If more information is stored for the user model, more precise the personalization of the system will be [4].

T. Ahram and W. Karwowski (eds.), *Advances in Human Factors, Software, and Systems Engineering*, Advances in Intelligent Systems and Computing 598,
DOI 10.1007/978-3-319-60011-6_8

Having in mind the concepts above, user modeling is a technique to define models for the system according with a set of user characteristics in a computational environment [1]. This process of user modeling analyzes and infer implicit information about the user [5]. The user models try to describe the cognitive process of user's actions, differentiate the expert skills of user's skills, it describes preferences or behavioral patterns, and the characteristics of the users [3]. User model can be built by the information provided by the user or in an automated process, which the creation of user model is controlled by the system, using learning techniques that make the model adaptive [5, 6].

Patterns in HCI are the project patterns with solutions of problems that commonly occurs, they help the designers of the system during the planning phase. The patterns have a vocabulary and specific structure to each field of study [7]. The interface pattern are patterns that help to organize the structure and behavior of the interface to build a more understandable and usable user interface [8, 9]. The interface patterns encourage the designer to employ established solutions to represent a recurrent problem and its abstract solution, so this solution can be applied to different instantiations of the same problem [10].

Human Interaction Robot (HRI) is a subfield of HCI, and it is the area of study applied in this research. Robot design should involve the user-centered development process considering the context usage. The interaction between people and robots must occur safely, using the robot as an intelligent instrument, adapting and anticipating the user needs.

HRI field studies how people behave while interacting with robots and it efforts to excerpt the greatest consequence from that. Alongside of how well a robot can help a user or how easy it can be used to complete a task, it should be considered how that user will respond while interacting with it [11].

The goal of this paper is to present a structure of interface patterns that allows future projects of adaptive interface and HRI to consider the automatic changes of the iterative process based on a set of patterns. This work presents some HCI concepts and methods related to human factors, adaptive interface and interface patterns, and finally, the pattern structure identified.

2 Human-Computer Interaction

The Human-Computer Interaction (HCI) is a multidisciplinary field of study that focus on the interaction between humans and computers through a "user interface" that intends to facilitate and increase human motivation, action and experience. The goal is to build systems that are more useful, usable and accessible to people [1, 12].

This section will present some important concepts of HCI, human factors, adaptive interface and interface patterns along with some elements that link the HCI field with the human factors.

2.1 Concepts and Human Factors

The definition of Human Factors (HF) or Ergonomics, adopted by the International Ergonomics Association (IEA) [13], is: "Ergonomics (or human factors) is the scientific discipline concerned with the understanding of interactions among humans and other elements of a system, and the profession that applies theory, principles, data and methods to design in order to optimize human well-being and overall system performance."

The HF study the interface physical aspects to attend the users' characteristics and how they perform their activities. The physical elements need to be taken into consideration on how to display the information, the control mechanism disposal, make the system comfortably reachable for all users, the color usage, and the health aspects like lightening, temperature, noise and time. This field contributes with HCI by identifying constraints of the system design process, and providing guidelines and standards [14, 15].

Psychology is one of the fields that has a strong relationship with HF, since it also concerns with the humans' perceptual limitations, and it contributes directly with HCI to understand humans' motivation, action and experience [12, 14]. The psychology study brings fundaments that are used to define some of the concepts and process of HCI considering the human aspects. The following concepts are just a few terms and techniques that represent the importance of the study of human aspects applied in HCI.

Gestalt Principles are used to fill blank spaces with elements in a graphic interface in order to make the application more efficient and accessible by displaying these elements in an easier way to visualize and interpret. Some of commonly known principles are: proximity, continuation, symmetry, similarity, common fate, closure, common region, uniform connectedness [16, 17].

"Semiotics" is the study of the signs, what they represent, and the process involved in the interpretation of these signs. A sign is anything that it is placed in some other element space, it contains particular aspects, indicating the thoughts to something else, like words, images, odors, flavors, acts or objects. [18]. Semiotic Engineering is described in HCI as the human communication through a computational system, focusing in the communication between designers, users and system [17].

The HF problems are associated with issues that impacts the system performance and the unexpected behavior of the system by not meeting with the user's expectations [19]. The following concepts and process are just a few of the terms used in HCI to minimize those problems.

The term "design", according with [14], can be partially defined as achieving goals through the constraints, which are material, standards, costs, etc. Design can be a verb, noun or an adjective, it can refer to the final product or a process, the design process may vary according with each situation and it represents a sequence of activities [12, 20]. The design is a process that encompasses the analysis and interpretation of the current situation; synthesis of a situation by planning and executing an intervention in the current situation; and evaluation of the new situation by comparing the old situation to the new acquired [17]. The "usability" is the quality of the system usage. It concerns and evaluate the effectiveness, efficiency and satisfaction of the user when executing tasks to achieve a goal [15]. According with Nielsen [21], Usability has numerous components but it is commonly associated with learnability, efficiency, memorability, errors and satisfaction.

The above concepts demonstrate the importance of understanding the user when building a system. To do so, it is necessary to collect data with information about the user, the user's relation to technology, knowledge of the system domain, knowledge of the tasks, and motivation and values. The user data is important to design appropriate system that accommodate the users' needs and it can be collected through interviews, questionnaires, focus groups, brainstorming, card classification, field study, contextual investigation. The data collected will be analyzed by the designer and disposed in representations or models, like user profile, personas, problem scenario, task models [17].

Human Interaction Robot (HRI) is a subfield of HCI that studies the interaction between the robot and humans, based on how people behave while interacting with robots and the greatest consequence from this interaction, how well a robot can help a user or how easy it can be used to complete a task [11]. It is a multidisciplinary field that considers human and robot factors, combining knowledge from HCI, psychology, cognitive science, social science, artificial intelligence, computer science, robotics and engineering [22].

2.2 Adaptive Interface

The adaptive system is a system that can adapt itself according with the user's goals and preferences, which are stored in a user model. The user model holds the information about the user's preferences in order to distinguishing one user from another. The role of user model can be described as the interpretation of user actions, and it provides accurate feedback according with the knowledge base [2].

The user model is also responsible for the adaptation inside the system through the internal actions, like filtering to receive a certain information [2]. It can be built with the information provided by the user or in an automated process, which the creation of user model is controlled by the system, using learning techniques that make the model adaptive. As more information are stored in the user model, more precise the personalization of the system will be [5].

Adaptive interface is an artifact with the objective to learn the user interactions to build user models and improve the system experience. The dynamic of the adaptive interface with the user occurs by mesmerizing the user's actions and enhancing the system interactions [3]. The adaptation process occurs according with the information retrieved about the user, information processed by the user model and the adaptation according with the user model [2].

There are some commonly known applications with adaptive user interfaces on internet that uses information for filtering and recommendation tasks. The first case is applicable to filter information to display what seems to be interesting and relevant for the user. The second case is used to recommend something to the user, like products or articles [3].

Adaptive techniques are used to improve system's performance, and it can be separate in two categories of adaptation. One, the systems provides the possibility to the user select and configure alternative presentation and interactions characteristics, like set preferences in a dialog, move and resize the user interface elements, etc. Second, the system should capture scenarios that requires adaptations and execute appropriate actions, monitoring user interactions and data, building assumptions to each user or group of users [23].

Some paradigms in adaptive interface are: the programming by demonstration, which constructs interfaces personalized according with the user actions; intelligent tutoring system, that personalize instructions; learning apprentices, created to solve complex problems using a knowledge-based, like learn an expert decision and replicate the same steps; there is a paradigm that builds a new model for each user by extracting user's preferences [3].

The universal access is a perspective that tries to capture and accommodate the user's abilities, experiences and needs of a product in a computer-based environment. It intends to design products that are adaptable attending a large number of users in different contexts. The Universal Access encompasses issues related to the content, functionalities and the interaction aspects of physical, semantic and syntactic characteristics. It is an automatic adaptation capable of capture, gather and attend the user needs, like presenting adaptive content to the users with different levels of experience regarding the application, attending user with or without disabilities [23].

2.3 Interface Patterns

Patterns can be defined as solution to recurring problems in a specific context, sharing project experiences and values that can be used by other professionals with lower experiences or similar problems. They can also be used as a communication channel and discussion of ideas [7, 24, 25].

Patterns is a set of knowledges containing description of the best practices in a specific domain, capturing common solutions in a project. They describe an element of the project with the reason of why should be used in a project, presenting the problem and the solution. A set of patterns contains several patterns that can be used isolated, a group of patterns that are under the same context generates a pattern language [7].

The pattern language contains its individuals' patterns with title, description of application argument, and a representation of patterns relationship (from the same language). It is a metalanguage to generate languages for problems in a domain. Each pattern from the pattern language should complete another pattern from the language, the first pattern creates a state that the next pattern will solve and together they all solve the overall problem [7, 25, 26].

Pattern Language Markup Language (PLML) was created in order to create more unified and accessible pattern language. PLML is a pattern language used to make the patterns more organized by having a more consistent terminology, and it allows the contradictions and failure recognition to be in a pattern language. The PLML was created to standardize the patterns. The Extended Pattern Language Markup Language (PLMLx) is a extended version of PLML, which contain the PLML element and also elements of authoring, management changes, and reference to other patterns. The required element of PLMLx ae: name, problem, context, forces and solution [7, 27].

User Interface Pattern are commonly known interface behaviors and solutions for problems with the intention to provide a practical guidance to the developers and designers, and to provide the user comfort and sense of familiarity when using the system [28, 29]. The interface pattern are patterns that help to organize the interface, such as structure and behavior, in order to build a more understandable and usable user interface, it increases the efficiency and consistency of the user interface, and it can also ensure good usability [8, 9].

3 Interface Patterns for Adaptive Interface

The increasing number of applications that need to attend a large number of users require the adaptation of interface in order to attend users' needs and the content adaptation according with the user's preferences. This section will present a pattern structure defined to attend system with adaptive interface.

As we saw in previous section, interface pattern helps to organize the interface to accommodate the user for the system usage, making the system more usable an accessible [8, 9]. The adaptive interface is used to build system that can adapt the content or the information display according with the user's needs and preferences [2]. The aspect of universal access incorporates the idea of adaptation to attend a broad diversity of people [23].

The intention to attend a large diversity of users can be challenging for the interface designer, who can face difficulties to identify the diversity in the solutions; the client, who can find the system difficult to be used; and the sponsor of the project, who might not achieve the desired objective in the system creation. Considering these issues, the use of patterns to document the recurrent problems can help to achieve positive usability, facilitate the communication between the designers, and standardize the solutions used in the project [7].

One way to design a project to attend different users is to map and understand the users' profile, which it can be represented by personas. Personas is the technique used to represent a group of users considering a set of representative variables that allow to characterize fictional characters. These characters should document behavioral information, ethnography and user needs [30].

In order to represent the diversity of users during the system development, [7] defined the Interface Patterns in Personas Layers (IPPL) as a set of patterns with solutions based on personas, or solutions in layers of significant representation of the users. The combination of personas and patterns supports the interface design with the representation of a set of users and patterns, documentation, team communication, and information of the problem context.

The IPPL are interface patterns that contain a structure organized in layers that are built and classified according with personas. The structure was based on required attributes from PLMLx, having the following basic structure [7]:

- Name: It defines the pattern associated with the problem and solution. The name should be a simple and relevant noun that links to the pattern details, and it should not be ambiguous with other patterns' name of the same set of patterns.
- Context: The context is the scenario of the problem with details of the situation and conditions that the issue occurs. It justifies the problem situation and reports the solution restrictions.
- Forces: It represents the tensions (contradictory aspects) involved in the problem, which are relevant considerations to choose the solution. They are implicit in the context description.
- Problem: It is the specific problem that the pattern intends to solve, and it should be described independent of the context.

- Persona: It is a name or an image defined for a group of users that represents the set of users identifies the persona layer.
- Solution: It contains the solution or several solutions for the problem according with the context, considering the forces and based on the persona. The solution should not describe a project or a specific interface, once the pattern is like a template to be applied in different situations. The solution can also contain an image to facilitate the representation but it should not influence in the interface creation process of the designer.

This structure allows to include several persona layers that describe personalized solutions according with each group of users. The Fig. 1 is a representation of the basic structure of IPPL.

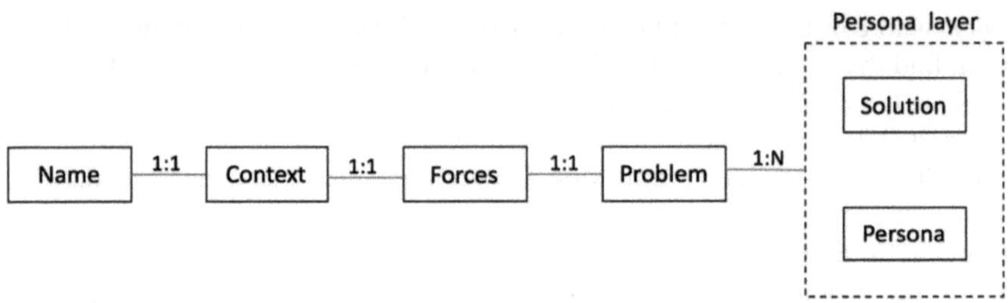

Fig. 1. Representation of the elements disposal from the IPPL and the respective cardinality.

The creation process of IPPL defines a sequence of ordered tasks to be performed by the designers in order to achieve the set of IPPL based on data that characterizes the group of users. It starts defining the personas and identifying the set of patterns. The personas are identified by collecting, organizing and storing user data with information of the user characterization, like personal data, computational experience and ethnographic data [7].

The user data can be gathered by an automatic process or user observation. The data mining can be applied to group similar information to identify the persona, which it will be created with a name, behavior description, needs and preference [7]. The study of [31] proposed an automated process to identify groups of users' profiles through a WEB interface based on the analysis of this interface usage. The process involves the user data collection, clustering by similarity, identify characterization, and analysis and definition of the profiles identified.

The personas are representations of the user profiles and each persona should have distinct characteristics from the other personas identified, in order to get an effective representation of the users' system. The set of patterns can be defined by choosing patterns stablished from literature (books, websites, papers) that are relevant for the project, or by identifying patterns specifically for the project, based on the interface usage or the process of documenting solutions [7].

After gathering the personas and set of patterns, the IPPL process analyzes the existing patterns solutions to define the solutions for the set of personas. The solution of each pattern must be related to the persona from the same pattern, and it must be based on the user behavior, patterns solutions from the literature, and usability testing. The set of IPPL need to be used during the entire software development lifecycle. The process of IPPL creation uses external artifacts (user data, interface patterns) and generates artifacts, like the resource that documents the set of patterns in personas layers. The flow of the process is performed by a usability specialist who executed the following activities: problem analysis based on the users' tasks; definition of the procedure to user data collection and storage structure according with the personas classification; user data analysis for personas characterization; identification of the set of patterns interface; personas creation; and IPPL definition [7].

The Fig. 2 is a representation of the result achieved from IPPL, containing one problem with two solutions that attends different personas.

Name		Map	
Context		The user need to perform a task to achieve his goal. The task consist on navigate through several level of interface to find what the user needs but the he does not know how to go back to the origin category.	
Forces		The user has the goal but he is not familiarized with the path to locate what he needs. To achieve a certain level of interface, the user needs to access its previous levels but he gave up from the path direction he was performing. To achieve the objective, several steps need to be taken but the number of steps may vary according with the decisions made during the navigation.	
Problem		The user achieves the goal after navigating though several levels of interface. The user gets lost and does not know how he got to that point of the interface. The user does not know how to go back using the application resources.	
Personas Layer	**Solution**	 According with the choices made by the user during the navigation, the interface can display only the path performed, so the user can check the current level and go back to the desired level. The user can visualize the set of levels that were accessed to achieve the current state. It is possible to go back N levels through the navigation map or breadcrumb.	 Always display all navigation options with a simplified representation of the paths available. Indicate the user current state in each moment of the navigation.
	Illustration		

Fig. 2. Adaptation of the IPPL example [7]

The first persona (on the on the left column of the Fig. 2) does not have too much time to learn new things besides what involves his job role, he has 42 years old, he uses the internet only to pay bills or read emails, he uses the computer only in his job, and he has low computational experience. The second persona (on the right column of the Fig. 2) likes to do things on his way, he has 26 years old, he is a postgraduate student, he uses to internet for most of his daily tasks, he uses the computer at home, work and school, he has high computational experience. Considering the divergence of the personas, different solutions were provided to the same problem in order to attend both group of users [7].

Typically, the Personas creation process is manual, lengthy and it doesn't have an attendance during the project. There are researches that presents a process to automatize and to address the user's needs through Personas during the whole project. They are important references that add up to the result of this research [32, 33]. They can be used as components for adaptation.

In order to obtain a structure of interface patterns that attends projects of adaptive interface and HRI with automatic changes of the iterative process based on a set of patterns.

The IPPL was adapted to have one more element: adaptation variables. This element will be responsible to store variables that represents the adaptation involved in each solution.

Some adaptation variables from a web environment can be extracted from paper [31], like: total time of the website navigation; navigation time through the website content; time that the user spent in questionaries' pages; average time used to fill text inputs of all forms; total time spent to fill non textual inputs of all forms; average time to fill intuitive text inputs of all forms; average of touch per second in text field of forms; average of touch per second in intuitive text fields of form; percentage of "backspace" usage for all text fields filled; percentage of "backspace" usage in intuitive text inputs; number of times that errors occurred to fill forms; number of times that the same error occurred to fill the form; number of repeated clicks in the same link; number of times that the user clicked in the help button; number of times that the user used the menu; usage of breadcrumbs links; "search" usage accessing the autocomplete items; number of times that the user clicked in links or buttons that are with focus on the main page; time of mouse movement over the website content; time of the mouse movement over the menu; number of user clicks outside buttons or links; duplicated access for the same content; time to visualize the initial page of the website; average time to visualize pages with detailed content; total time of the visualization of an informative page about research; total time to visualize the main page.

The adaptation variables above were collected according with the users' interactions in order to identify the profiles automatically based on the navigation characteristics. The list of variables are some examples in a website environment and they should be adapted according with each project.

With the example of Fig. 2, these adaptation variables could be used for both personas to capture the time that the users take to go back to a specific point, the total time of navigation, repeated actions that leaded to the same page, etc. But some different variables could be used to estimate the time of usage for each case like counting the click of the breadcrumb for the first persona and the clicks on the menu for the second persona, or the time that the user spent with the mouse over the menu. The values obtained for the adaptation variables can help to identify a new profile, for example, if the time to go back to a specific step or the time (or number of click) that the user spent trying to use the menu are too divergent, maybe we are dealing with two different groups of users, like one that has experience with technology and the other not. The variables can also help to improve the solution and usability. If the time to complete a task is too long, if the user spent too much time to take an action to the correct direction or if the number of erroneous click is too high, maybe the solution provided does not attend the problem to that persona, so the solution can be improved, including more intuitive aspects.

The interdisciplinary field of HRI studies the interaction between the human and robot, focusing on functionalities and usability aspects to design and implement robots and the tasks that involves humans [34, 35]. So for a human-robot interaction project, the interface is the robot itself, in this case some of the variables could be: time to approach the user, speed to approach the user, voice volume, voice speed, robot expression, time of the user response, number of times that the user executed the expected feedback, number of erroneous response, total time to achieve the goal, number of repeated actions, etc.

For the structure of the interface patterns of this paper, the idea is to combine the adaptation variable with IPPL, including this element in the personas layer. So, the variables should be included according with each persona and its respective solution, providing a better understanding of the persona, improving usability and supporting (or even help to improve) the solution of the problem.

Using the same personas from Fig. 2, the following problem was used as an example in a HRI environment to represent IPPL combined to adaptation variable. The problem focus on what is the best expression that the robot should demonstrate when approaching the human to communicate a failure. The Fig. 3 is a representation of a water leak problem, which the robot tries to communicate the user to raise the ecological awareness and reduce waste of water. Considering the difference of the personas, the robot can use two different approaches. The persona with low experience requires a soft approach to avoid frustration or negative sentiments to this technology. For the persona with more familiarity to technology, the robot can have a stronger expression to emphasized the problem.

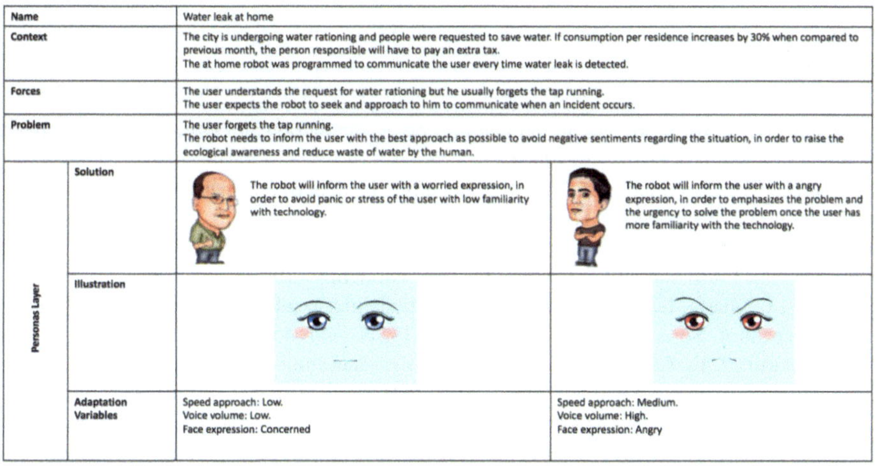

Name		Water leak at home	
Context		The city is undergoing water rationing and people were requested to save water. If consumption per residence increases by 30% when compared to previous month, the person responsible will have to pay an extra tax. The at home robot was programmed to communicate the user every time water leak is detected.	
Forces		The user understands the request for water rationing but he usually forgets the tap running. The user expects the robot to seek and approach to him to communicate when an incident occurs.	
Problem		The user forgets the tap running. The robot needs to inform the user with the best approach as possible to avoid negative sentiments regarding the situation, in order to raise the ecological awareness and reduce waste of water by the human.	
Personas Layer	**Solution**	The robot will inform the user with a worried expression, in order to avoid panic or stress of the user with low familiarity with technology.	The robot will inform the user with a angry expression, in order to emphasizes the problem and the urgency to solve the problem once the user has more familiarity with the technology.
	Illustration		
	Adaptation Variables	Speed approach: Low. Voice volume: Low. Face expression: Concerned	Speed approach: Medium. Voice volume: High. Face expression: Angry

Fig. 3. The IPPL combined the adaptation variables for a HRI scenario.

Considering that IPPL should be used during the entire software development lifecycle, the adaptation variables can be constantly updated according with the interface generated and the usability tests. The results and numbers collected for each variable provide an average to the designer, who can use those values to improve the solutions and the system usability during the software lifecycle.

4 Conclusion

The IPPL can be a fundamental resource to support the diversity of users in projects with adaptive interface, obtaining patterns by mapping the personas and providing personalized solutions for each persona in the persona layer. In order to obtain a set of patterns that consider automatic changes of the iterative process, the IPPL was adapted to include the adaptation variables in the persona layer to help the designers to improve the usability of the personalized solutions, and even identify new users' profiles based on the interaction behaviors.

The new structure of interface pattern defined in this paper (IPPL combined with adaptation variables) can be used not only for computer interface with adaptive content but also in other fields such as human-robot interaction. The interaction with the robot is something that requires adaptation, once different people might have different action and response. The elements of the structure of interface patterns proposed in this paper need to be adapted according with the project, context and problem. The patterns using this structure can help to map the variables of adaptation, the personas, and the personalized solutions to attend a large number of scenarios.

Acknowledgments. This work is supported by FEI (Fundação Educacional Inaciana Pe. Sabóia de Medeiros) and FAPESP (Fundação de Amparo à Pesquisa do Estado de São Paulo), process number 2017/04942-1.

References

1. Fischer, D.: User modeling in human-computer interaction. J. User Model. User-Adap. Interact. (UMUAI) **11**, 65–86 (2001)
2. Fröhlich, C.: User modeling and user profiling in adaptive e-learning systems. Master's Thesis, Graz University of Technology (2005)
3. Langley, P.: User modeling in adaptive interfaces. Adaptive system group. In: Proceeding of the Seventh International Conference on User Modeling (1999)
4. Martins, C., Faria, L., Carrapatoso, E.: User modeling in adaptive hypermedia educational. Educ. Technol. Soc. **11**(1), 194–207 (2008). (ISSN-1436-4522)
5. Frias-Martinez, E., Chen, S.Y., Liu, X.: Survey of data mining approaches to user modeling for adaptive hypermedia. IEEE Trans. Syst. Man Cybern. Part C Appl. Rev. **36**, 734–749 (2006)
6. Aquino Junior, P.T., Filgueiras, L.V.L.: User modeling with personas. In: Proceedings of the 2005 Latin American Conference on Human-Computer Interaction (CLIHC 2005), pp. 277–282. ACM, New York (2015). doi:https://doi.org/10.1145/1111360.1111388
7. Aquino Jr., P.T.: PICaP: padrões e personas para expressão da diversidade de usuários no projeto de interação. Escola Politécnica, University of São Paulo, São Paulo, Brazil. Doctoral Thesis in Sistemas Digitais (2008). doi:10.11606/T.3.2008.tde-15092008-144412. Accessed 08 Feb 2017
8. Zhao, M., Gao, Y., Liu, C.: Research and achievement of UI patterns and presentation layer framework. In: Fourth International Conference on Computational Intelligence and Communication Networks (2012)
9. Rossi, G., Schwabe, D., Lyard, F.: User interface patterns for hypermedia application. In: AVI 2000 Proceedings of the Working Conference on Advanced Visual Interface, pp. 163–142 (2000)
10. Aquino Junior, P.T., Filgueiras, L.V.L.: A expressão da diversidade de usuários no projeto de interação com padrões e personas. In: VIII Brazilian Symposium on Human Factors in Computing Systems (IHC 2008), pp. 1–10. S. B. Computação, Porto Alegre (2008)

11. dos Santos, T.F., de Castro, D.G., Masiero, A.A., Aquino Junior, P.T.: Behavioral persona for human-robot interaction: a study based on pet robot. In: Kurosu, M. (ed.) HCI 2014. LNCS, vol. 8511, pp. 687–696. Springer, Cham (2014)
12. Carroll, J.M.: Designing Interaction: Psychology at the Human-Computer Interface. Cambridge University Press, Cambridge (1991)
13. International Ergonomics Association. http://www.iea.cc/whats/index.html
14. Dix, A., Finlay, J., Abowd, G.D., Beale, R.: Human-Computer Interaction. Pearson, Upper Saddle River (1993)
15. Cybis, W., Betiol, A.H., Faust, R.: Ergonomia e Usabilidade: Conhecimentos. Métodos e Aplicações, Novatec (2007)
16. Ogden, C.K.: Principles of Gestalt Psychology. Routledge, Abingdon (1935)
17. Barbosa, S.D.J., Silva, B.S.: Interação Humano- Computador. Ed CAMPUS (2011)
18. Chandler, D.: Semiotics. Routledge, Abingdon (2002)
19. Stanton, N.A., Salmon, P.M., Rafferty, L.A., Walker, G.H., Baber, C., Jenkins, D.P.: Human Factors Methods: A Practical Guide for Engineering and Design. Ashgate, Farnham (2013)
20. Lawson, B.: How Designers Think: The Design Process Demystified. Elsevier Fourth Edition (2005)
21. Nielsen, J.: Usability Engineering. Morgan Kaufmann, Burlington (1993)
22. Dautenhahn, K., Saunders, J.: New Frontiers in Human-Robot Interaction. Advances in Interaction Studies 2. John Benjamins Publishing Company, Amsterdam (2011)
23. Stephanidis, C.: Adaptive techniques for universal access. User Model. User-Adap. Interact. **11**, 159–179 (2001)
24. Seffah, A., Desmarais, M.C., Metzker E.: HCI, usability and software engineering integration: present and future, vol. 8, pp. 37–57. Springer (2005)
25. Todd, E.G., Kemp, E.A., Phillips, C.P.: Introducing students to UI patterns. In: CHINZ 2009 Proceedings of the 10th International Conference NZ Chapter of the ACM's Special Interest Group on Human-Computer Interaction, pp. 37–40 (2009)
26. Kaisler, S.H.: Software Paradigms. Wiley Interscience, Hoboken (2005)
27. Hennipman, E.-J., Oppelaar, E.-J., van der Veer, G.: Pattern languages as tool for discount usability engineering. In: Graham, T.C.N., Palanque, P. (eds.) International Workshop on Design, Specification, and Verification of Interactive Systems, pp. 108–120. Springer, Heidelberg (2008)
28. Sacramento, C., Paiva, A.C.R.: Web application model generation through reverse engineering and UI pattern inferring. In: 9th International Conference on the Quality of Information and Communications Technology (2014)
29. Thanh-Diane, N., Vanderdonckt, J., Seffah, A.: UIPLML: pattern-based engineering of user interfaces of multi-platform systems. In: 2016 IEEE Tenth International Conference on Research Challenges in Information Science (RCIS) (2016)
30. Filgueiras, L., Aquino Junior, P.T., Sakai, R., Filho, A.G., Torres, C., Barbarian I.: Personas como modelo de usuários de serviços de governo eletrônico. In: Proceedings of the 2005 Latin American Conference on Human-computer interaction (CLIHC 2005), pp. 319–324. ACM, New York (2005). doi:https://doi.org/10.1145/1111360.1111395
31. D´Ângelo, F. M.: Identificação automática de perfis de grupos de usuários de interfaces web. Master's Thesis, University Center of FEI (2012)
32. Masiero, A.A., Tonidandel, F., Aquino Junior, P.T.: Similar or not similar: this is a parameter question. In: Yamamoto, S. (ed.) Human Interface and the Management of Information. Information and Interaction Design, HIMI 2013. LNCS, vol. 8016, pp. 484–493. Springer, Heidelberg (2013)

33. Masiero, A.A., Carvalho Destro, R., Curioni, O.A., Aquino Junior, P.T.: Automa-persona: a process to extract knowledge automatic for improving personas. In: Stephanidis, C. (ed.) HCI International 2013 - Posters' Extended Abstracts, HCI 2013. Communications in Computer and Information Science, vol. 373, pp. 61–64. Springer, Heidelberg (2013)
34. Mutle, B.: Interaction with robotic technologies. In: HRI Japan, South Korea, and China, p. 59. World Technology Evaluation Center, Arlington, (2012)
35. Feil-Seifer, D., Matarié, M.J.: Human-Robot Interaction (2012)

A Stochastic Programming Model for Resource Allocation with Uncertainties and Carbon Emission Constraints

Wen Yang, Yanchun Pan(✉), and Xingyu Chen

College of Management, Shenzhen University, Shenzhen, China
{wenyang, panyc, celine}@szu.edu.cn

Abstract. Cap-and-Trade mechanism has been implemented in many countries. To study the resource allocation problem with limited carbon emission, a stochastic dynamic programing model is proposed. The company produces two types of products to satisfy random demands with uncertain supply. One is normal product with higher revenue and emission, and the other is green product with lower revenue and emission. The optimal allocation policy is characterized, which is a target type with a threshold. The numerical example illustrate the analytical results and indicates that the cap-and-trade system as a market-based mechanism can induce the companies to produce green products.

Keywords: Resource allocation · Carbon emission constraints · Cap-and-trade · Stochastic dynamic programming · Optimization

1 Introduction

The globe is experiencing the effects of climate change. Human activities are largely responsible for climate change and other global environment problems, primarily due to the greenhouse gas emission produced by energy consumption during production and service. For the purpose of environmental protection, many legislations or systems are established, i.e. the European Union Emissions Trading System (EU-ETS). China sets the target of 40–50% emissions reduction (from the 2005 baseline) by 2020 [1]. To meet this goal, China started pilot emission cap and trade system in 2011 in seven cities and plans to launch a national system in 2017. In the system, the companies are allocated an amount of emission permit to offset their pollutant emissions. Excess emission results in high penalty. The total amount of emission allowance is controlled by a pre-determined cap, which is gradually tightened over time.

As the main subject of carbon emission, the manufacturing companies are facing great challenges under this system. In order to reduce their emissions, companies can resort to improving production processes to produce green products with lower profit, for instance, using less polluting machines or cleaner materials. On the other hand, the companies would like to satisfy customers' various demands (i.e. product (green or normal), quantity) as much as possible. The company needs to decide whether to accept or reject an incoming customer order based on the available supply as well as the constrained emission allowance to avoid from penalty. Moreover, there exist

T. Ahram and W. Karwowski (eds.), *Advances in Human Factors, Software, and Systems Engineering*, Advances in Intelligent Systems and Computing 598,
DOI 10.1007/978-3-319-60011-6_9

uncertainties in the supply chain. For example, the demands and replenishment of materials usually deviate from the forecast and scheduled supply. These uncertainties make the order acceptance/rejection decision even more complicated. The traditional and practical order acceptance decision in this situation is to satisfy the incoming orders as much as possible without reserving resources (supply and emission permit) for the future orders. However, this method is not benefit to the profit. In addition, the previous research on order acceptance decision solely determined by demand and supply may not be optimal any more when carbon emission is constrained.

This study considers an order acceptance problem with constrained carbon emission and uncertain demand and supply in a finite time horizon. The company provides two kinds of products: green product with lower profit and unit emission, the other one is normal product with higher profit and unit emission. Accepting the orders requiring normal product will earn more profit but also emit more. On the contrary, accepting the orders with green products results in less profit but lower carbon emission, which may lead the company to avoid penalty and, what's more, earn additional revenue by selling the surplus emission permit in the carbon trading market at the end of the horizon. The objective of this study is to maximize the expected net profit over the planning horizon by determining which order is going to be fulfilled.

The remainder of the paper is organized as follows. Section 2 reviews the relevant literature. Section 3 describes the problem being considered and presents the model. Section 4 analyze the structural properties of the model. Section 5 describes a numerical example. Finally, Sect. 6 concludes the work.

2 Literature Review

For a comprehensive overview on the demand management problems, see the review works [2]. Most of the papers study the demand fulfilment in a deterministic condition, where all the orders during the planning horizon are known, and thus the order acceptance decision is to select the most profitable orders [3–6].

However, managers will not be able, naturally, to obtain all the necessary information concerning future orders when making decisions. Some papers propose stochastic demand management model to allocate resources to ensure optimal promising of uncertain customer orders. Spengler *et al.* formulate a stochastic model for the iron and steel industry characterized by multiple resources with capacity constraints to promise different classes of uncertain demands [7]. Gao *et al.* concern with committing two types of resources in assemble-to-order (ATO) system, including production capacity and component inventory, to multiclass orders with different unit revenues and constant delivery lead times [8]. However, they do not consider the emission constraints.

Another stream of related literature is production-planning problem with constrained CO2 emission. Absi *et al.* propose four lot-sizing models integrating carbon emission constraints. These models help decide the optimal production period, place and quantity [9]. Helmrich *et al.* study a lot-sizing problem with an emission constraint under concave nonlinear cost and emission functions [10]. Chen *et al.* investigate optimal order quantities based on EOQ model under various regulations [11]. Manikas and Kroes investigate the newsvendor problem under cap-and-trade and carbon tax policies and get the optimal production quantity and profit [12].

To our best knowledge, few studies have investigated a resource allocation problem, considering multiple periods, multiple products, limited carbon emission, and uncertain demand and supply simultaneously. This paper is going to propose a stochastic model to solve this problem.

3 Model Formulation

At the beginning of the compliance period (i.e. one year), the government allocates an emission permit allowance to the manufacturing company, say Q. At the end of this horizon, if the actual total emission is larger than Q, the company will be punished. If any surplus, the emission permit can be sold to earn extra profit.

Under the carbon emission control, in order to reduce emission, the manufacturing company can improve production process to produce green products will less unit emission, u_g. However, the green products will cause higher production cost and thus bring lower unit revenue, r_g. Besides the green product, the company also provides normal product with more unit emission, $u_n (u_n > u_g)$, and higher unit revenue $r_n (r_n > r_g)$. This paper studies how to make the resource allocation decision on orders from different customer classes. In this study, the customer orders are divided into two classes c: one is requiring green product, g, the other is normal product, n. And the probability of two kinds of orders are P_g and P_n respectively, where $P_g + P_n = 1, g, n \in c$. The order quantity and supply of material in each period, d_t and s_t, is independent random variables. At the beginning of period t, the inventory of material and emission permit level is I_t and E_t respectively. After the company decides how much of the order should be fulfilled, the inventory level reduces to Y_t. In other words, the accepted quantity is $I_t - Y_t$. The unit holding cost of unused material is h. Suppose $e\bar{u} < \bar{r} + h$, where $\bar{u} = u_g P_g + u_n P_n$, $\bar{r} = r_g P_g + r_n P_n$. It means that the expected revenue of selling the emission permit is less than that of selling a product. Otherwise, the company will not organize the production activities and sell all the emission permit instead.

The objective is to maximize the total profit of all the products in the compliance horizon and a dynamic programming model is formulated. The maximum expected profit from period t to T, $V_t(I_t, E_t)$, is defined as

$$\begin{aligned} V_t(I_t, E_t) &= \max_{Y_t}\left\{\mathbb{E}_{d_t,c}\left[r_{c,t}(I_t - Y_t) - hY_t\right] + \mathbb{E}_{s_t}\left[V_{t+1}(Y_t + s_t, E_{t+1})\right]\right\} \\ &= \mathbb{E}_{d_t,c}\left[r_{c,t} I_t\right] + \max_{Y_t}\left\{\mathbb{E}_{d_t,c}\left[-(r_{c,t} + h)Y_t\right] + \mathbb{E}_{s_t}\left[V_{t+1}(Y_t + s_t, E_{t+1})\right]\right\} \end{aligned} \tag{1}$$

where

$$E_{t+1} = E_t - (I_t - Y_t)u_{c,t}, E_1 = Q \quad c = g, n \tag{2}$$

$$(I_t - d_t)^+ \le Y_t \le I_t \tag{3}$$

$$I_t \in [I_{t,min}, I_{t,max}] \tag{4}$$

For simplicity, suppose $V_{T+1}(I_{T+1}, E_{T+1}) = eE_{T+1}$, where e is the carbon price. It implies that if there is surplus emission permit by the end of the horizon, the company can sell it in the trading market.

4 Structural Properties

This section will analyze the structural properties of the model formulated above. First, define

$$Z_t(Y_t) = \left[-\left(r_{c,t}+h\right)Y_t\right] + \mathbb{E}_{s_t}\left[V_{t+1}(Y_t+s_t, E_t-(I_t-Y_t)u_{c,t})\right]. \quad (5)$$

Hence, (1) becomes

$$V_t(I_t, E_t) = \mathbb{E}_{d_t,c}\left[r_{c,t}I_t\right] + \mathbb{E}_{d_t,c}[\max_{Y_t}\{Z_t(Y_t)\}]. \quad (6)$$

In the following theorem, the convexity of the optimal profit functions is established.

4.1 Theorem 1

If the terminal function $V_{T+1}(I_{T+1}, E_{T+1})$ is a convex function, $V_t(I_t, E_t)$ and $Z_t(Y_t)$ are both convex functions.

Proof. If $V_{T+1}(I_{T+1}, E_{T+1})$ is convex, then $V_{T+1}(Y_T+s_T, E_T-(I_T-Y_T)u_{c,T})$ is a convex function in Y_T, where I_T is a state variable here. Therefore, $Z_T(Y_T)$, which is a decreasing function $-\left(r_{c,T}+h\right)Y_T$ plus a convex function $\mathbb{E}_{s_t}[V_{T+1}(Y_T+s_T, E_T-(I_T-Y_T)u_{c,T})]$, is still convex. Since the function $\max_{Y_T}$ and $\mathbb{E}_{d_T,c}$ will not change the convexity, $\mathbb{E}_{d_T,c}[\max_{Y_T}\{Z_T(Y_T)\}]$ is convex. Then $V_T(I_T, E_T)$ is concave. Following this logic, $V_t(I_t, E_t)$ and $Z_t(Y_t)$ are convex in each period.

Thus, based on Theorem 1, it is investigated that the optimal policy which is shown in the next theorem.

4.2 Theorem 2

For $t = 1, \ldots, T$, for any given state (I_t, E_t) at the beginning of period t, there exist threshold S_t, such that optimal policies are characterized as follows:

(a) *Suppose $I_t \le S_t$. Then, $Y_t^* = I_t$, it means the optimal policy is to reject the incoming customer order.*
(b) *Suppose $I_t > S_t$. Then, $Y_t^* = \max\{S_t, I_t - \min\{d_t, \frac{E_t}{u_{c,T}}\}\}$. This implies that the company should satisfy the customer order and the satisfied quantity equals to $I_t - Y_t^* = \min\{I_t - S_t, d_t, E_t/u_{c,T}\}$.*

Proof. By using Theorem 1, it is found that there exists an optimal solution, S_t, to $Z_t(Y_t)$, which is proved to be a convex function. If $Z_t(Y_t)$ is a decreasing function, then $S_t = I_{t,min}$. If $Z_t(Y_t)$ is an increasing function, $S_t = I_{t,max}$. If $Z_t(Y_t)$ is strictly convex, then $S_t \in (I_{t,min}, I_{t,max})$. The optimal policy is to satisfy the customer order to the base level S_t to maximize the expected profit in each period.

5 Numerical Example

In this section, a numerical example is given to demonstrate the preceding analytical results. The planning horizon is $T = 5$. The replenishment of raw material in each period following discrete uniform *DU* (0, 1) distribution. The unit holding cost of material is $h = \$1$. The unit revenue of normal and green products are $r_n = \$50$, $r_g = \$30$. The unit emission of producing two products are $u_n = 2$, $u_g = 1$. The order arrival probabilities for two products are equal, i.e. 1/2. The order quantity follows discrete uniform distribution *DU* (0, 9). The carbon price is $e = \$25$. A simulation procedure is implemented to generate the random number and obtain the expected value.

By using the dynamic programming method, the thresholds of two products in each period are obtained. Figure 1 shows the results when the surplus emission permit can be sold or not in the trading market at the end of the planning horizon.

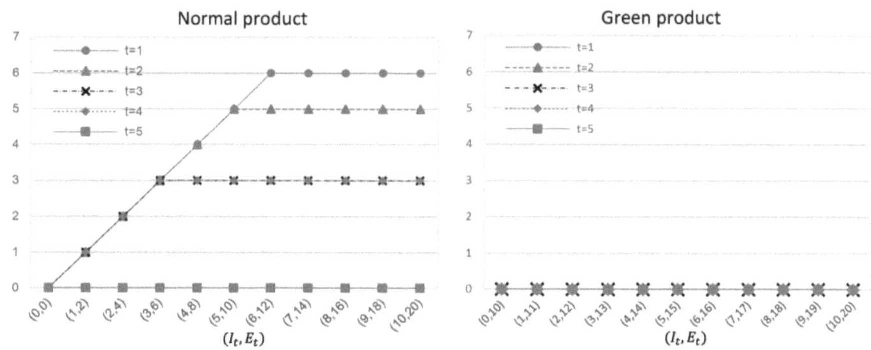

(a) The surplus emission permit can be sold

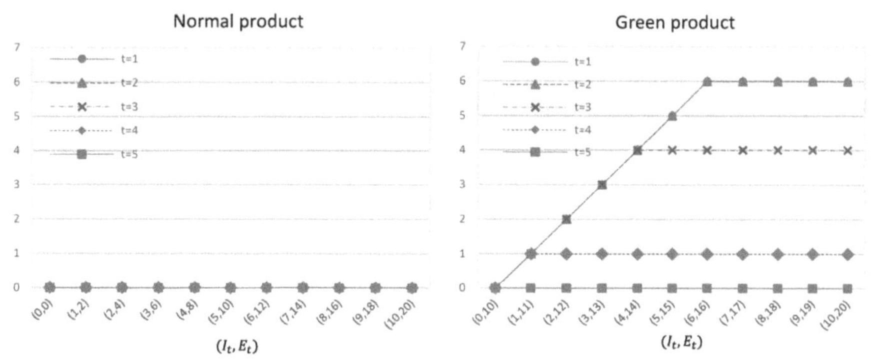

(b) The surplus emission permit can not be sold

Fig. 1. The optimal thresholds of normal and green products in two scenarios

From Fig. 1, it is found that the thresholds are different in each period and increase to a level with increasing (I_t, E_t). In period 1, for instance, when the materials and emission permit (I_5, E_5) are not available to satisfy 6 normal products, the threshold is increase and the incoming order should be reject. Otherwise, the order should be accepted-but only up to the threshold 6. For example, if the state is $(I_5, E_5) = (10, 20)$ and the order quantity is 8, then the order is partially accepted, only 4 is satisfied. If the order quantity is less than 4, the order should totally accepted.

In addition, it is known that the threshold decreasing with increasing period. Take the normal product in Fig. 1(a) for example, the thresholds of period 1 and period 5 are 6 and 0 respectively. It is because if the surplus emission permit can be sold, although the unit revenue of the green product is lower, the saved emission permit can earn more profit than the normal product. Thus, in the early period, the company should reserve the resources for the future potential demand for the green product to earn more profit. But for the green product, the threshold is zero in each period, which means that the order acceptance policy is to satisfy the customer order as much as possible. However, when the surplus emission permit can not be sold (Fig. 1(b)), the result is opposite. The company prefer to reserve the resources to the normal product whose unit revenue is higher. Therefore, the cap-and-trade scheme, under which the emission permits are tradable in the market, can incent the companies to produce green products which results in emission reduction. And higher carbon price can induce the companies to invest on the green technology.

6 Conclusion

Under the cap-and-trade system, where the total emission is controlled, the manufacturing companies are facing more challenge on making production decisions. This study is oriented to a resource allocation problem with constrained carbon emission and uncertain demand and supply. The company provides more profitable and polluting normal product as well as less profitable and less polluting green product. The surplus emission permits are tradable in the trading market at the end of the planning horizon. A stochastic dynamic programming model is proposed for the problem. It is characterized the optimal policy that maximizes the expected the total profit over a finite planning horizon. In particular, the optimal policy is a target level type with a threshold decreasing with the period. The numerical example is further conducted to illustrate the analytical results and the advantage of the cap-and-trade system in emission reduction.

It is meaningful to develop an efficient heuristic algorithm to obtain the thresholds. In addition, in this study, the product type is selected by the customers. In some situation, it is the company to decide which product should be delivered to the customer. This is left to be the future research.

Acknowledgments. The work described in this paper was supported by Research Grant from the Natural Science Foundation of China (NSFC) under grant codes 71402102, 71472126, 71572114 and 71402101.

References

1. Zhou, M., Pan, Y., Chen, Z., Chen, X.: Environmental resource planning under cap-and-trade: models for optimization. J. Clean. Prod. **112**, 1582–1590 (2016)
2. Kilger, C., Meyr, H.: Demand fulfilment and ATP. In: Stadtler, H., Kilger, C. (eds.) Supply Chain Management and Advanced Planning, 4th edn, pp. 181–198. Springer, Heidelberg (2008)
3. Slotnick, S.A., Morton, T.E.: Selecting jobs for a heavily loaded shop with lateness penalties. Comput. Oper. Res. **23**(2), 131–140 (1996)
4. Christou, I.T., Ponis, S.: A hierarchical system for effective coordination of available-to-promise logic mechanisms. Int. J. Prod. Res. **47**(11), 3063–3078 (2009)
5. Lin, J.T., Hong, I.H., Wu, C.H., Wang, K.S.: A model for batch available-to-promise in order fulfillment processes for TFT-LCD production chains. Comput. Ind. Eng. **59**(4), 720–729 (2010)
6. Jung, H.: An available-to-promise model considering customer priority and variance of penalty costs. Int. J. Adv. Manuf. Technol. **49**(1–4), 369–377 (2010)
7. Spengler, T., Rehkopf, S., Volling, T.: Revenue management in make-to-order manufacturing-an application to the iron and steel industry. OR Spectr. **29**(1), 157–171 (2007)
8. Gao, L., Xu, S.H., Ball, M.O.: Managing an available-to-promise assembly system with dynamic short-term pseudo-order forecast. Manag. Sci. **58**(4), 770–790 (2012)
9. Absi, N., Dauzere-Peres, S., Kedad-Sidhoum, S., Penz, B., Rapine, C.: Lot sizing with carbon emission constraints. Eur. J. Oper. Res. **227**, 55–61 (2013)
10. Helmrich, M.J.R., Jans, R., van den Heuvel, W., Wagelmans, A.P.M.: The economic lot-sizing problem with an emission capacity constraint. Eur. J. Oper. Res. **241**, 50–62 (2015)
11. Chen, X., Benjaafar, S., Elomri, A.: The carbon-constrained EOQ. Oper. Res. Lett. **41**, 172–179 (2013)
12. Manikas, A.S., Kroes, J.R.: A newsvendor approach to compliance and production under cap and trade emissions regulation. Int. J. Prod. Econ. **159**, 274–284 (2014)

Body Temperature Monitoring System for Slaughterhouse Workers

Saulo Güths, Victor Augusto dos Santos, Fabiano Takeda, Diogo Cunha dos Reis, and Antônio Renato Pereira Moro(✉)

Universidade Federal de Santa Catarina, Campus Universitário Trindade, Florianópolis, SC 88040-900, Brazil
saulo@lmpt.ufsc.br, viic.santos@gmail.com, takeda.f@bol.com.br, diogo.biomecanica@gmail.com, renato.moro@ufsc.br

Abstract. The aim of this study was to develop a complete system to evaluate the thermal stress in slaughterhouses. The proposed system consists of a wireless sensor network (WSN) using Zigbee technology, providing a set of sensor nodes to measure the slaughterhouse workers body temperature in different regions simultaneously and a another set to measure the environment conditions. The data are collected by a USB Zeegbee module connected to a Windows PC and the PMV (Predict Mean Vote) is showed at real time. A test was conducted in the sector of cuts of a poultry slaughterhouse industry located in the south of Brazil. The results showed a virtually thermal comfort of the workers (by PMV analysis), but a dangerous low values which may cause disease to the worker.

Keywords: Poultry slaughterhouse · Body temperature monitoring system · Thermal stress evaluation

1 Introduction

One of the strongest economic activities of the south Brazil region is the chicken production for exporting. World rankings point to Brazil as the largest exporter of frozen poultry meat since 2004 and the third producer after USA and China [1, 2]. Most of the Brazilian exported chicken meat is in the form of cuts (54.7%), and the remainder consists of processed meat (4.6%), salted meat (4.5%), and whole chicken (36.2%) [1]. In Brazil, the poultry farming employs more than 3.6 million people, mostly being line workers. Due to the lack of machinery providing full automation in the sector, most of the chicken meat produced in Brazil is processed through manual labor, using of hand tools live knives and chainsaws. The frequently exposure to an artificially cold environment combined with the poor working conditions lead the slaughterhouse workers to develop a variety of cold related injuries [3].

According to Caso et al. [4], the slaughterhouse workers are exposed to biomechanical risk factors for development of work-related musculoskeletal disorders of the upper limbs (UL-WMSDs) such as repetitiveness, high-frequency technical actions, excessive use of force, inappropriate postures, insufficient recovery time, handling of frozen products and prolonged exposure to the cold environment.

T. Ahram and W. Karwowski (eds.), *Advances in Human Factors, Software, and Systems Engineering*, Advances in Intelligent Systems and Computing 598,
DOI 10.1007/978-3-319-60011-6_10

Through their studies in different Brazilian slaughterhouses, Tirloni et al. [5] and Reis et al. [6] reported that the majority of workers felt discomfort during their working day in a variety of body regions, such as shoulders, neck, spine, arms, hands and wrists.

The aim of this study was to develop a complete system to evaluate the thermal stress in slaughterhouses, evaluating both the environment conditions and the thermal discomfort of the workers.

2 Market Research

The present study started with a market research, in order to look for the existence of a product or complete solution capable of monitor the environment conditions as well as tracking the worker body temperature in real-time. The company 3 M™ has a product line dedicated to thermal stress analysis named QUESTemp™. QUESTemp™ II is a personal heat stress monitor kit capable to tracking the worker's in-ear temperature minute-by-minute. QUESTemp™ 36 is a heat stress monitor data logging kit, which offers logging of environment conditions in real time such as dry bulb, wet bulb and globe temperatures along with relative humidity and air speed.

Although interesting, both solutions are very expensive [7, 8] and neither have the capacity of tracking the temperature of different body parts of the worker, which is the key characteristic present in the proposed system. In fact, the in-ear temperature is an important measurement to identify thermal discomfort, but the anatomical regions associated with the most prevalent WMSDs are extremities such as the wrist and hand, and epidemiological research supports the evidence of injuries related to repetitive and forceful hand-intensive tasks [9].

3 Hardware and Software

3.1 Introduction

The proposed system consists of a wireless sensor network (WSN) using Zigbee technology, in which a set of sensor nodes is intended for the measurement of the slaughterhouse workers body temperature in different body regions simultaneously, while another sensor node evaluates the environment conditions, such as relative humidity, air speed, globe temperature and dry bulb temperature.

Both nodes send the sampled data in real-time over a radio link to a radio sink node connected to a computer through a USB-serial converter (Fig. 1). All the collected information is centralized in a friendly interface (Window system), which presents the data to the user. Also, the system should be non-intrusive and easily expansible.

3.2 Zigbee Network Technology

The wireless sensor network (WSN) implemented in this project is entirely based on the Zigbee wireless technology, using Xbee-Pro 900 XSC S3B modules manufactured by Digi International, Inc. Zigbee communication is specially built for control and

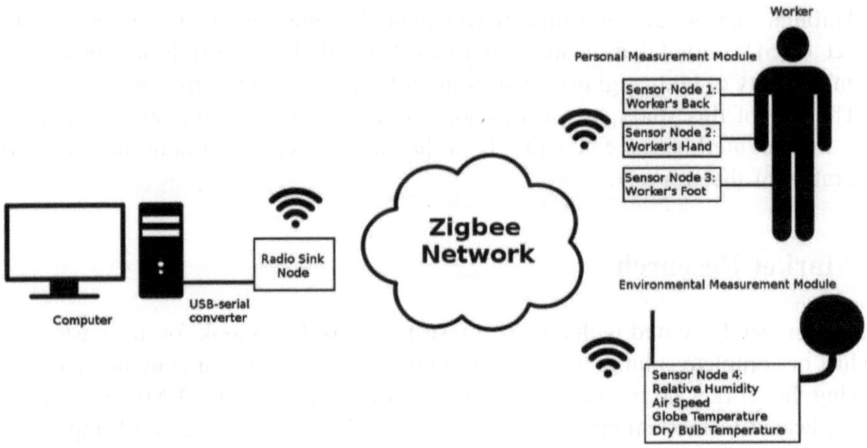

Fig. 1. Implemented system to evaluate thermal stress in slaughterhouses.

sensor networks based upon IEEE 802.15.4 standard for wireless personal area networks (WPANs). The main reasons of using Zigbee network technology and Xbee modules are the small size, low power consumption, low cost of the modules, long battery life, simplicity to setup the network, long range and the built-in peripherals such as 4 10-bit ADC inputs, 2 PWM outputs, Serial interface (UART) and up to 15 digital I/O ports. Xbee module works with a power supply range from 2.1 to 3.6 VDC, and it's line of sight range can reach up to 15 km with appropriate antenna and configurable Data Rate. In terms of network topology, the implemented network was structured in the Star Topology, where all the network sensing nodes communicates directly to only one sink node, which is connected to the computer.

3.3 Personal Measurement Module

The Personal Measurement Module was configured to measure the slaughterhouse workers body temperature in different body regions simultaneously. Each module (Fig. 2) consists of a battery charger, a one-cell Lithium-Ion battery and a low-dropout DC-DC converter used to drop-down the 3.7 Vdc from the battery to the power supply range of the Xbee module.

The thermistors used are Murata Thermo String NTC thermistors, with the B-Constant of 3380 $\pm$ 1% and 10 k $\pm$ 1% Ohm resistance (at 25 °C). The "head" has a diameter equal to 1.2 mm, the wire has 0.30 mm diameter and length 80 mm. These thermistors are characteristic to provide an extremely precise temperature sensing, excellent long-term stability and high malleability. The sensors have been soldered in a flexible AWG 26 double wire plastic insulated

Each thermistor is connected to a Xbee ADC input through a voltage divider with a 10 K $\pm$ 1% Ohm precision resistor. A precision voltage regulator supplies a reference voltage of 1.2 Vdc to the sensors, to match with the internal Xbee ADC voltage reference, in order to obtain a full-scale conversion. The ADC sample rate is 4 Hz.

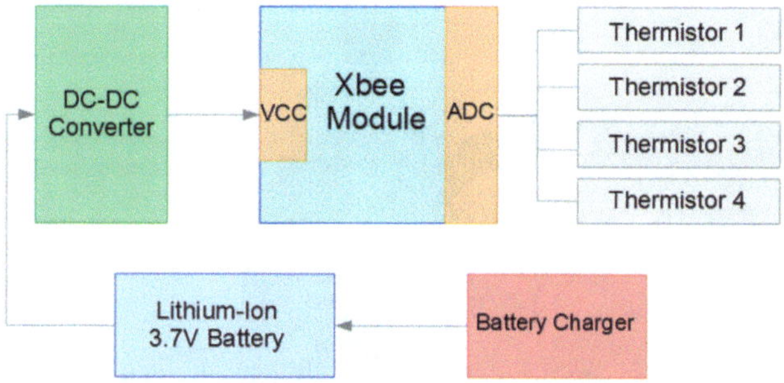

Fig. 2. Implemented system to evaluate thermal stress in slaughterhouses.

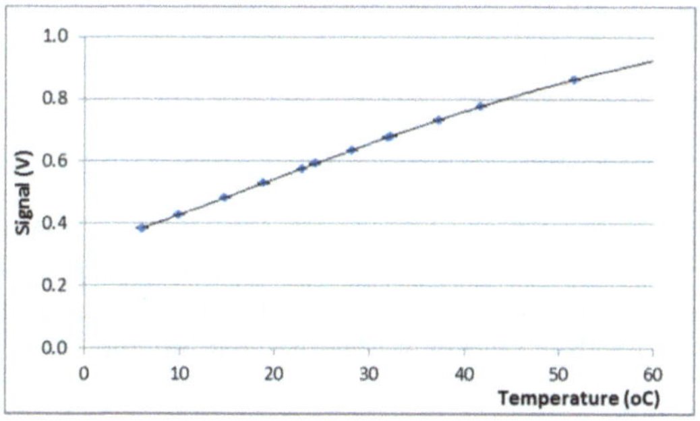

Fig. 3. Typical thermistor calibration curve.

Each thermistor was individually calibrated in water (thermostatic bath) using as reference a thermometer TESTO 735-1 (uncertainty equals to 0.01 °C). A fourth-order polynomial calibration curve was generated for each sensor with a R^2 better than to 0.999 (Fig. 3). The maximum combined error is 0.1 °C in the 5 to 40 °C range.

It was used three Surface Temperature Module per worker attached to the hand, foot and back (Fig. 4). The thermistors have been fixed with micropore tape directly on the worker's skin (Fig. 5). The gloves and socks were put on directly over the sensors. The ear temperature sensor was placed around a silicone earplug.

3.4 Environmental Measurement Module

The Environmental Measurement Module has a very similar block diagram to the Personal Measurement Module, differing in the type of battery (a lead acid 7.5 V battery) and the charger. The relative humidity is measured by a Honeywell HIH 5030 sensor, which provides a high level linear voltage output, very low drift, low power consumption and uncertainty around 1%. The sensor was connected directly to the Xbee module ADC input. The air temperature is measured by a thermistor similar to described at Sect. 3.3. The thermistor is installed inside a forced ventilated tube (Fig. 8) to prevent error due the radiant exchanges. A micro-ventilator (20 mm diameter) is positioned downstream.

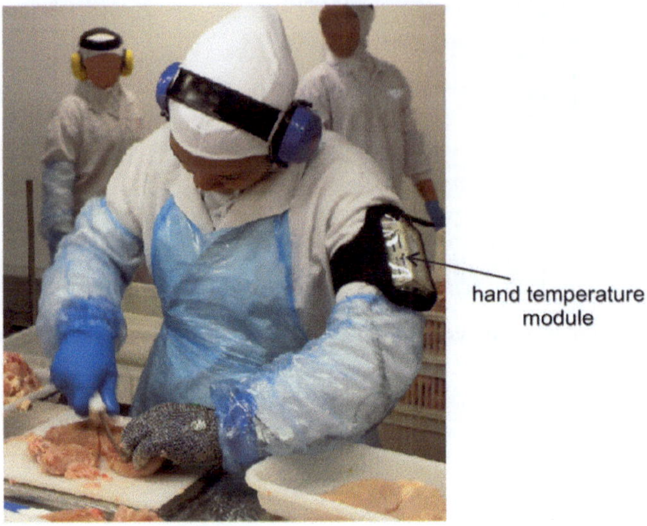

Fig. 4. Worker (Participating staff freely consented to the research protocol, which was submitted and approved by the Ethics Committee of the Federal University of Santa Catarina and fulfilled all legal recommendations. Anonymity and confidentiality of information were maintained in data record) with the Personal Measurement Module.

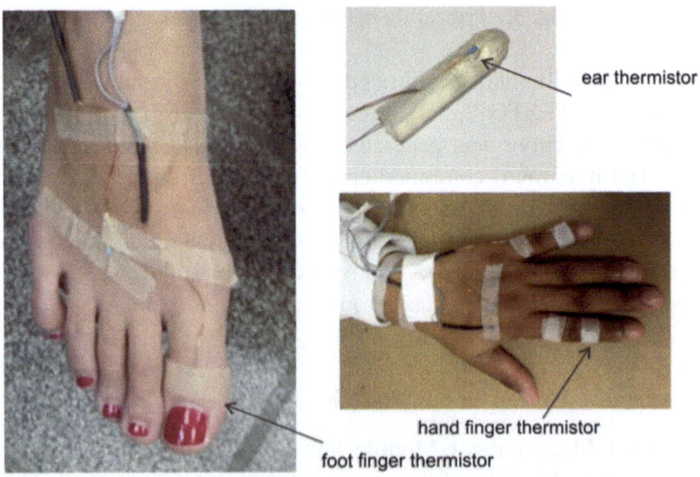

Fig. 5. Surface temperature sensors.

The air speed is measured by a thermo-anemometer made by a thermistor heated by a constant heat source. A Constantan wire (0.1 mm diameter 30 mm long), has been wrapped in the thermistor and a constant 3.3 VDC was applied. A metalic shining cap has been glued onto the assembly as show at Fig. 6.

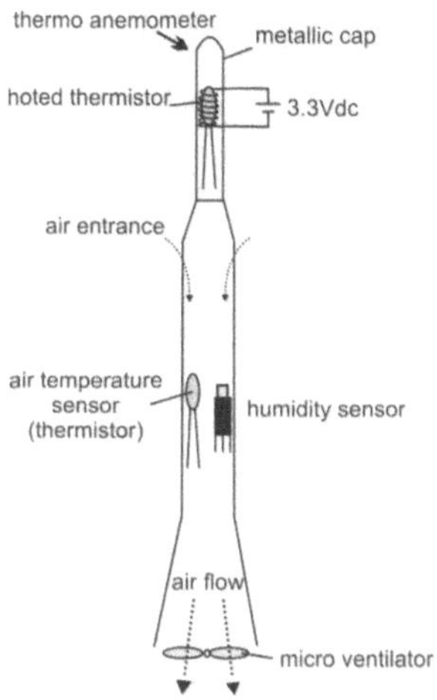

Fig. 6. Details of the anemometer, humidity and air temperature sensors.

The thermistor heats aproximately 18 K above de room temperature at zero air speed. Increasing the air speed the hoted thermistor temperature decrease. The temperature difference (Delta_T) between the hoted thermistor and the air temperature sensor is the output signal of this anemometer. The Fig. 7 shows the calibration curve using as reference a TESTO 435 anemometer, (uncertainty equals to 0.01 m/s) as reference in a wind tunel. The combined incertitude is 1% + 0.02 m/s.

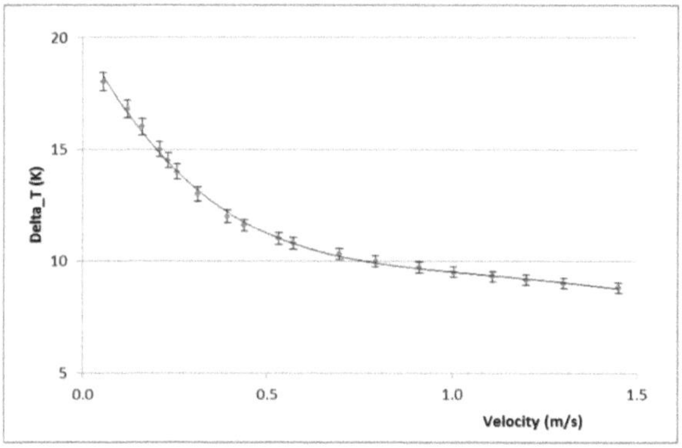

Fig. 7. Thermo anemometer calibration curve.

The globe temperature measurement consists of a hollow aluminium sphere of 150 mm diameter (thickness 1 mm) painted matte black (IR emissivity 0,93) with a temperature sensor (thermistor) at its centre. Figure 8 shows the prototype of the Environmental Measurement Module.

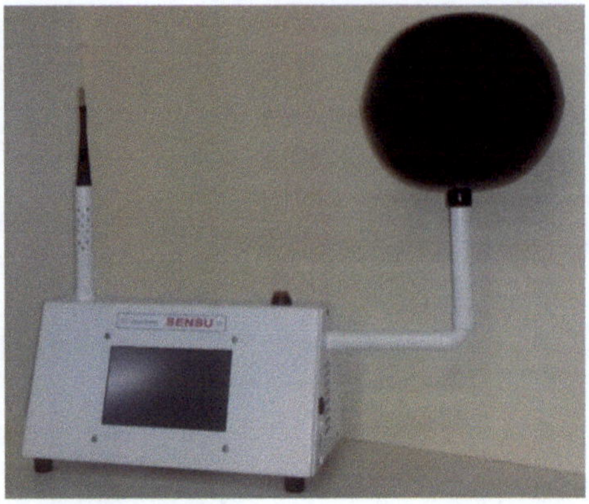

Fig. 8. Prototype of the environmental measurement module.

With these sensor is possible to determine de PMV (Predicted Mean Vote) and PPD (Predicted Percentage Dissatisfied). These indexes, defined by ISO 7730 and ASHRAE 55 based on Fanger studies in 1970, predicts the mean comfort response of a group of people practicing an activity and wearing a certain clothing. However this study does not take account of the discomfort caused by thermal assimetrie of the environment or the contact with hot or cold surfaces.

3.5 User Interface

The graphical user interface was developed in C++ language using Borland C++ Builder IDE for Microsoft Windows. All data received goes through a moving average filter to remove the noise and are also show to the user through a graphical interface. A post processing software allow modify the metabolic rate and clothing index (Fig. 9).

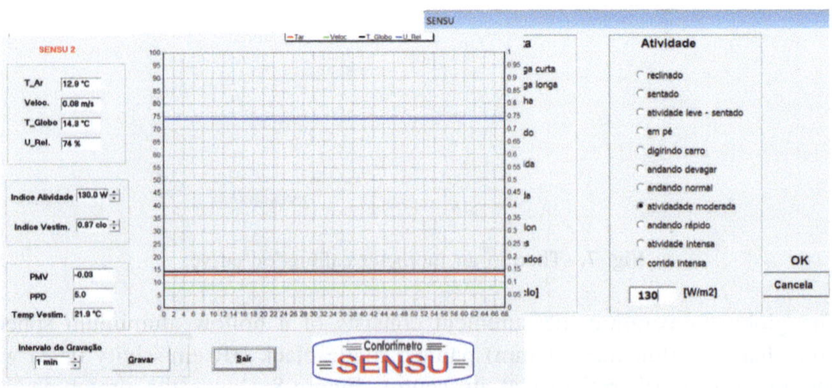

Fig. 9. User Interface.

4 Results and Discussion

A test was conducted in the sector of cuts of a poultry slaughterhouse industry located in the south of Brazil. A worker was instrumented with three surface temperature modules connected in his left hand (hand that holds the product), on his right foot and on his back. He was using a latex glove (0.15 mm thickness) overlaid with a cotton glove (1.5 mm thickness) and also a steel mesh glove (2.5 mm thickness) in the hand holding the product. He was wearing a PVC boot (4.0 mm thickness) and cotton sock (2.2 mm thickness).

The Environmental Measurement Module has been placed near to the worker and the recorded values are showed at Fig. 10-a. It can observed that the air temperature reach a minimal temperature equal 12.5 °C, not considered unhealthy by the Brazilian worker protection law.

a)

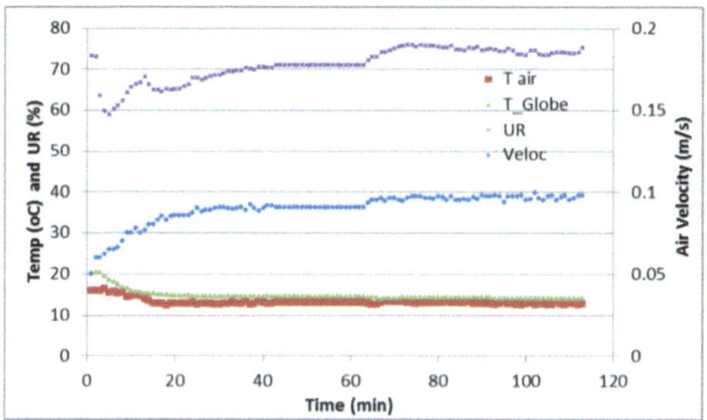

b)

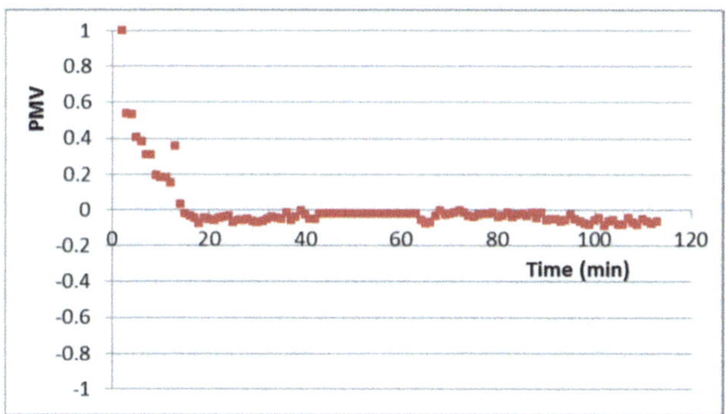

Fig. 10. (a) Environmental data, and (b) PMV values.

The PMV (Predict Mean Vote) is showed at Fig. 10-b, considering a metabolism rate equal 130 W/m^2 and a clothing index equal 0.87. These values have been respectively obtained by ISO 8996/04 for this kind of activity and by ISO 9920/0 considering the usual clothes of the workers. It can be observed that the PMV is close to zero which means that the worker are in thermal equilibrium and theoretically in thermal comfort condition. But the Fig. 11 shows the low temperature of the hand finger and the foot finger. The hand finger temperature reached 16 °C for certain period and the foot finger temperature has been between 17 °C and 20 °C during the major part of the period. Kaminski et al. [13] and Ilmarinen e Tammela [14] have been realized study in slaughtering industry where the finger temperatures were quite similar. They showed that this condition can induce health problem, particularly the Raynaud's Disease.

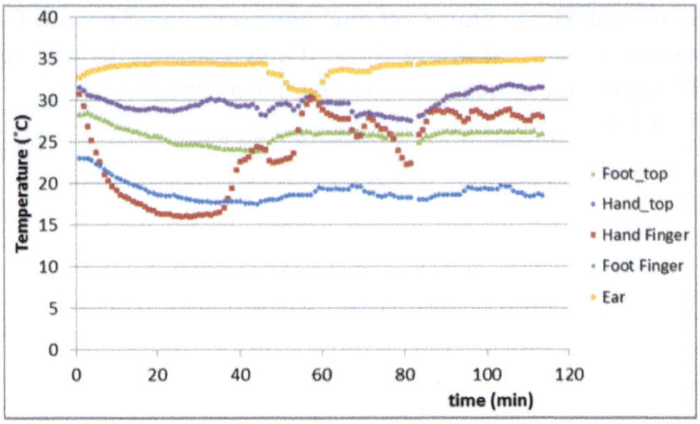

Fig. 11. Surface body temperature.

5 Conclusion

The proposed system met the design requirements, being able to acquire the worker's temperature with a precision of 0.1 °C, due to the meticulous procedure of calibrating the thermistors. Moreover, the arrangement of thermistors did not disturb the activity of the worker. The sensor responsible for acquiring environmental conditions data maintained the same reading accuracy of temperature, relative humidity and air speed data. The construction of the anemometer showed simple and it has a robustness suitable for an industrial application. The presented system has been proved as an easy and reliable way to Ergonomists and Security Engineers to evaluate the environment conditions and thermal stress inside slaughterhouses.

The tests performed in an industrial slaughterhouse showed despite an adequate PMV and a lawful air temperature, the hand and food finger temperature reach dangerous low values which may cause disease to the worker.

References

1. Brazilian Poultry Association, 2015 Poultry Industry. http://www.brazilianchicken.com.br
2. Chicken Exports by Country. http://www.worldstopexports.com/chicken-exports-by-country/

3. Piedrahita, H.: Perception of musculoskeletal symptoms in cold-exposed and non-cold exposed workers. Department of Human Work Sciences, Division of Industrial Ergonomics, pp. 1–56 (2003)
4. Caso, M.A., Ravaioli, M., Veneri, L.: Esposizione a sovraccarico biomeccanico degli arti superiori: la valutazione del rischio lavorativo neimacelli avicoli. Prevenzione Oggi **3**(4), 9–21 (2007)
5. Tirloni, A.S., Reis, D.C., Santos, J.B., Reis, P.F., Barbosa, A., Moro, A.R.P.: Body discomfort in poultry slaughterhouse workers. Work **41**, 2420–2425 (2012)
6. Reis, P.F., Peres, L.S., Tirloni, A.S., Reis, D.C., Estrázulas, J.A., Rossato, M., Moro, A.R.P.: Influence of anthropometry on meat-packing plant workers: an approach to the shoulder joint. Work **41**, 4612–4617 (2012). Reading, MA
7. 3M Waterless Wet Bulb QUESTemp Area Heat Stress Monitor. http://www.raeco.com/acatalog/Buy-3M-Waterless-Wet-Bulb-QUESTemp-Area-Heat-Stress-Monitor.html
8. 3M QUESTemp II Personal Heat Stress Monitor. http://www.raeco.com/acatalog/Buy-3M-QUESTemp-II-Personal-Heat-Stress-Monitor.html
9. Viikari-Juntura, E., Silverstein, B.: Role of physical load factors in carpal tunnel syndrome. Scand. J. Work Environ. Health **25**(3), 163–185 (1999)
10. VCL Overview. http://docwiki.embarcadero.com/RADStudio/Berlin/en/VCL_Overview
11. What is Zigbee Technology, Architecture and Its Applications? https://www.elprocus.com/what-is-zigbee-technology-architecture-and-its-applications/
12. Lehmskallio, E., Hassi, J., Ketunen, P.: The skin in the cold. Int. J. Circumpolar Health **61**, 277–286 (2002)
13. Kaminski, M., et al.: Risk factors for Raynaud's phenomenon among workers in poutry slaughterhouses and canning factories. Int. J. Epidemiol. **26**(2), 371–380 (1997)
14. Ilmarinen, R.E., Tammela, E.K.: Design of functional work clothing for meat-cutters. Appl. Ergon. **21**(1), 2–6 (1990)

Applications

Efficient Language Model Generation Algorithm for Mobile Voice Commands

Daniel Yaeger(✉) and Christopher Bubeck

US Army, Aberdeen Proving Ground, USA
daniel.r.yaeger.ctr@mail.mil

Abstract. The Single Multimodal Android Service for HCI (SMASH) framework implements an automated language data generation algorithm to support high-accuracy, efficient, always-listening voice command recognition using the Carnegie Mellon University (CMU) PocketSphinx n-gram speech recognizer. SMASH injects additional language data into the language model generation process to augment the orthographies extracted from the input voice command grammar. This additional data allows for a larger variety of potential outcomes, and greater phonetic distance between outcomes, within the generated language model, resulting in more consistent probability scores for in-grammar utterances, and fewer false positives from out-of-grammar (OOG) utterances.

Keywords: Automatic Speech Recognition · ASR · Human computer interaction · HCI · Mobile computing

1 Overview

In support of an "always listening" configuration for Automatic Speech Recognition (ASR), in which the user's voice is the primary mode of interaction with the software system, there are three common approaches: The "wakeup word" approach, popularized by Apple (Siri) and Amazon (Echo), the keyword spotting from continuous speech approach, and the trailing silence approach (SMASH). The wakeup word approach, while effective for single interactions, adds an additional step to each interaction, making it cumbersome for continuous system interaction over time. Keyword spotting, which allows for instant voice interaction without an explicit action from the user, carries an increased risk of false positives when used in a continuously listening configuration. The trailing silence approach, implemented by SMASH, provides the structured interaction pattern of the wakeup word configuration while greatly reducing the amount of time necessary to indicate an intended interaction to the system.

In order to achieve a useful "always listening" voice command implementation, the underlying speech recognizer must provide quick response, high accuracy for in-grammar utterances, and a low false-positive rate for utterances that are not intended as voice commands. The latter is of utmost importance since the vast majority of speech received by the recognizer in an always-listening configuration will consist of utterances outside the command list. The system must be as effective at determining when speech is not a valid command as it is at recognizing the commands themselves.

T. Ahram and W. Karwowski (eds.), *Advances in Human Factors, Software, and Systems Engineering*, Advances in Intelligent Systems and Computing 598,
DOI 10.1007/978-3-319-60011-6_11

Grammar-based speech recognizers are typically used for voice command applications due to their ability to limit the potential results to a list of valid commands. The defined grammar, along with a Confidence Score, which communicates the recognizer's level of certainty in its result, and a minimum confidence threshold that filters out the more uncertain results, provide a solid foundation for an always-listening voice command implementation. Grammar-based recognizers are common in embedded applications, like those found in car stereos or ATMs, where system memory and processing power are markedly limited.

Initial versions of SMASH utilized the Nuance VoCon Hybrid speech recognizer, which provided a grammar-based recognizer with trailing silence detection and confidence score calculation. The recognizer provided satisfactory performance within SMASH, returning results with high accuracy and low rates of false positives when used with a reasonable minimum confidence threshold. However, in the interest of minimizing software lifecycle costs, licensed products, like Nuance VoCon, were removed from SMASH.

For the most recent version, the SMASH team began working with Carnegie Mellon University's PocketSphinx open-source speech recognizer as a potential replacement for the Nuance VoCon recognizer. While PocketSphinx supports grammar-based recognition, it does not calculate a confidence score for its results. Since always listening operation is not feasible without the inclusion of a useful confidence score, the grammar-based functionality in PocketSphinx was unsuitable for use in SMASH.

However, the PocketSphinx language model-based n-gram recognizer does calculate a probability score, which is provided along with its recognition results. The n-gram recognizer is designed to provide speech recognition of spontaneous speech, using a language model to provide a recognition result for any utterance. This type of recognizer is typically used for speech dictation applications, but with some additions and modifications can be used to handle voice command recognition as well. Necessary modifications include using trailing silence detection as an end point for recognition, using a grammar map to link results to a list of acceptable commands, and generating a language model using a voice command grammar as the basis for training data. This combination produces a recognizer that is capable of fast response and high accuracy for utterances within the expected voice command grammar. However, this configuration results in probability scores that do not track logically with utterances that are out-of-grammar (OOG), resulting in a high rate of false-positives when run continuously. An additional step is required in the language model generation process to allow the recognizer to provide usable probability scores and to minimize false positives.

The Single Multimodal Android Service for HCI (SMASH) 4.0 framework implements an automated language data generation algorithm to support high-accuracy, efficient, always-listening voice command recognition using the Carnegie Mellon University (CMU) PocketSphinx n-gram speech recognizer. SMASH injects additional language data into the language model generation process to augment the orthographies extracted from the input voice command grammar. This additional data allows for a larger variety of potential outcomes, and greater phonetic distance between outcomes, within the generated language model, resulting in more consistent probability scores for in-grammar utterances, and fewer false positives from out-of-grammar utterances.

2 SMASH Voice Command Workflow

SMASH provides voice command functionality to Android applications through a service-based approach, using a well-defined API set. SMASH performs all of the system level activities required to capture and recognize speech, providing speech recognition results to applications as Android intents [1]. The defined list of voice commands, or grammar, is defined by the integrating application, and is provided to SMASH at runtime using the SMASHBNF format. SMASHBNF is a specialized context-free grammar format based on the open-source JSpeech Grammar Format (JSGF) [2]. SMASHBNF allows developers to define the valid voice commands for their application, and link multiple commands to single actions using numeric ID tags.

SMASH utilizes the CMU PocketSphinx n-gram language model-based speech recognizer [3] for all speech recognition tasks, including both voice commands and dictation. Since this particular recognizer is not designed to handle grammar-defined recognition, multiple steps must be taken before the recognizer can begin listening for an application's specific set of voice commands. These steps are covered in detail in the remaining sections of this document. For the purposes of this section, and from the user's perspective, SMASH accepts a grammar from the application and begins to listen for the voice commands defined within.

Once SMASH begins listening, it analyzes the incoming audio for speech. Once speech is detected, recognition begins, and continues until 600 ms of silence occurs. In this case, "silence" is defined as audio signal that does not contain speech. Once the 600 ms of silence is detected, the recognizer provides a result and an associated confidence score. If this result matches a phrase from the grammar, and the confidence score meets the minimum threshold set by the integrating application, SMASH broadcasts an Android intent containing the result and its associated numeric ID. Once a result is generated, SMASH immediately resumes listening for incoming speech, beginning the process all over again.

The resulting workflow allows a user to speak any command from the grammar at any time, preceded and followed by a short pause (600 ms default value). Since the commands need to be separated from normal speech to match those in the grammar, normal continuous speech is typically not misrecognized as a command. Through the brief pause, the user is effectively indicating their intention to interact with the system. This is a similar workflow to a wakeup word configuration, but removes the extra call and answer between the user and the system, resulting in reduced time to interaction. This advantage is especially prevalent when executing multiple commands in a row.

In the following scenarios, a user interacts with a voice-enabled application to access information about the weather via both audible response and a display. In the first scenario, the system employs a wakeup word, "Alright system", to initiate a voice interaction.

"Alright system."

<Audible Confirmation>

"What's the weather like today?"

"Cloudy, 65 degrees, with a chance of rain."

"Alright system."

<Audible Confirmation>

"Show me the precipitation graph."

<Audible Confirmation>
<System Displays Weather Graph>

"Alright system."

<Audible Confirmation>

"What's the forecast for Memphis?"

"Sunny, with a high of 72."

In the second scenario, the system employs a trailing silence-based workflow.

<Brief Pause> "What's the weather like today?"
"Cloudy, 65 degrees, with a chance of rain."
<Brief Pause> "Show me the precipitation graph."
<Audible Confirmation>
<System Displays Weather Graph>
<Brief Pause> "What's the forecast for Memphis?"
"Sunny, with a high of 72."

In both scenarios, the user accesses the same three pieces of information using only their voice. In the first scenario, the system requires the user to speak six separate phrases to access the information. The user must also wait for audible confirmation for each wakeup phrase before speaking the desired command. In the second scenario, the user speaks three phrases, one for each piece of information, cutting the number of required interactions in half for the same result. In a simple scenario like this, the additional interactions might seem like only a slight inconvenience. However, in scenarios that involve a large number of continuous system interactions, or that are time-sensitive in nature, the cumulative additional time required for these extra interactions can seriously affect the user's ability to operate the system effectively. Given the advantages in efficiency, maintaining a trailing silence-based workflow was a priority during the development of SMASH 4.0.

3 Enabling Always Listening Voice Command Operation in SMASH with the CMU PocketSphinx Speech Recognizer

As previously noted, multiple operations must occur within SMASH to allow the CMU PocketSphinx language model-based n-gram speech recognizer to function in the same manner as one that is grammar-based. The following sections describe these operations in more detail to illustrate the challenges that were overcome using the tools made available by CMU on the way to creating a more robust, and more accessible version of SMASH.

3.1 Always Listening Voice Command Service

SMASH runs as an Android service that is contained in an Android Package (.apk) and an Android archive (.aar) for installation on devices and integration into apps. The main function of SMASH is to manage the connections between the speech recognition engine and the implementing applications. SMASH contains the Carnegie Mellon University (CMU) PocketSphinx speech recognition engine and acoustic model, allowing it to natively recognize spoken words and process them using a grammar-based voice command approach. Because the engine loads the grammar, language model, and dictionary (as appropriate) at runtime, the service itself does not need to house any of these components. This allows a single.apk to handle any application's ASR requirements without the need to recompile and redeploy SMASH for each new integration. The.aar library contains all the API calls and configurations necessary to use SMASH within applications.

For recognizing voice commands, SMASH achieves an observed "always listening" recognition capability by adding both fixed and dynamic break points in the loop based upon the incoming audio signal. Upon initiation of the engine, the recognizer listens to the incoming audio stream for possible speech input. When speech is detected, it begins trying to decipher speech from the incoming audio stream. At this point, the listener continues to try to decipher speech until the loop times out (configurable value, SMASH default is 120 k ms), or silence is detected (configurable value, SMASH default is 600 ms). When one of these conditions is reached, the recognizer provides its best hypothesis consisting of an orthography, or formatted text result, which is then matched to the current grammar to find an ID. If an ID is found, SMASH sends out a message containing the ID, the orthography, and a confidence score. The confidence score is represented by a positive integer value, with one being the lowest confidence. We have found that, depending on the environment and grammar, confidence score thresholds that consistently reflect an actionable item are generally in the 4700–5300 range. Once the recognizer has provided its results, the system begins listening for audio once again and the loop continues.

An Android application can bind to the SMASH service as it would any other Android service [4]. This includes implementing the Android ServiceConnection interface. The ServiceConnection interface notifies the client when the Android runtime has connected or disconnected the SMASH service from the client application. Next, the client app should register an Android BroadcastReceiver with the runtime, so that notifications can be received and processed by the app.

To start the recognizer, the SMASH service provides a startListening(…) method. The method parameters here include the SMASH BNF grammar, language model, and dictionary file, which provide the input for what speech should be listened for. The startListening(…) method is what actually starts the CMU speech recognition engine, and triggers the intent generation process when valid speech is recognized *(see Notification Generation section).*

3.2 Grammar Graph Generation

SMASH uses modified versions of the tools in the CMU Sphinx4 toolkit to build a directed graph of nodes that represent a SMASHBNF grammar [5]. From the Sphinx4 docs, "After the grammar is read in, it is converted to a graph of words representing the grammar. Let's call this the grammar graph. It is from this grammar graph that the eventual search structure used for speech recognition is built."

We will use the following simple grammar to generate an example grammar graph:

```
#SMASHBNF V1.0;
grammar example;
<SIMPLE_COMMANDS> = (
(microphone | mike) check {1} |
testing the (microphone {1} | mike {1}) |
trying out voice commands {2}
);
public <commands> = <SIMPLE_COMMANDS>;
```

When SMASH starts listening to a grammar, it generates the grammar graph for the input grammar. The graph for our example grammar looks like this (Fig. 1):

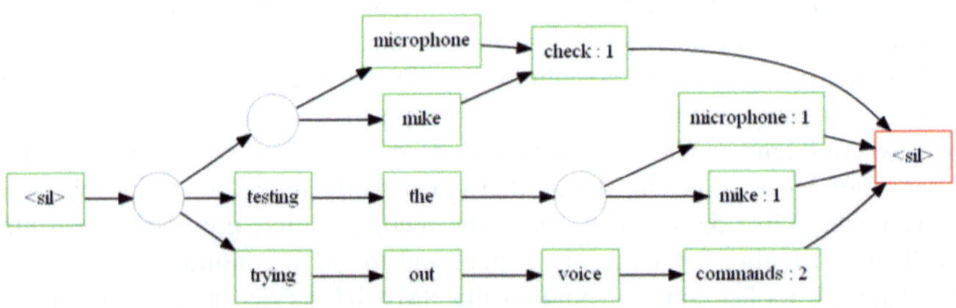

Fig. 1. Grammar graph

We have modified the Sphinx4 code to have the graph nodes keep track of SMASH IDs. These IDs are represented by tags '{}' in the grammar, and with a colon (e.g. 'check: 1'). When the PocketSphinx recognizer generates a recognition result, it returns an orthography, and a probability score. SMASH then searches the grammar graph for nodes that contain an ID value, and that match the orthography provided by the recognizer. The search algorithm used is an optimized depth-first search, which prunes the graph immediately when it is not possible for the orthography to exist in that part of the graph. Before searching, SMASH adds starting silence and trailing silence tags (<sil>) to the orthography. The orthography is also split by any whitespace to generate a list of tokens.

For example, when the orthography "testing the mike" is returned from the recognizer, SMASH would then search for "<sil> testing the mike <sil>". The first token, "<sil>", is the first node in the graph, so the search can continue. The next level of nodes to search are now [microphone], [mike], [testing], and [trying] (we pass through any empty nodes). We can now prune out [microphone], [mike], and [trying], since the search orthography cannot be in that part of the graph. The search continues with the [testing] node, and continues in the same way. By the end of the search, we will have the nodes [<sil>], [testing], [the], [mike: 1], and [<sil>]. Since "1" is the only ID in this path, we know that this orthography exists in the grammar, and can now notify the SMASH service client. Multiple IDs should not exist in any given path, if the grammar was written correctly. If this does occur, SMASH uses the ID closest to the final node in the graph.

Conversely, if an orthography is not found in the grammar graph, or there is no ID associated with an orthography, no notification to the client is generated.

3.3 Notification Generation

SMASH notifies service clients of a recognition by broadcasting an Android intent [1]. The intent contains the SMASH ID, modified probability score, orthography, and modality (e.g. VOICE_COMMAND). Once the client application receives the intent, it then executes the appropriate action linked to the specific ID received. The use of numeric IDs allows multiple orthographies, typically with the same meaning, to be linked to a single software action without requiring the client app to parse the orthography result itself, simplifying the voice command integration process.

4 Automated Model Optimization Data (AMOD)

To provide a language model that includes all of the phrases from a specific grammar, an existing language model must be retrained, or adapted, using the list of potential phrases in the grammar; or the grammar's orthographies can be used to train a completely new language model. Both approaches each have their own advantages, as well as potential pitfalls.

General use language models are designed to provide a basic model for recognizing spontaneous speech, which requires wide coverage of potentially tens of thousands of words and millions of phrases. However, these generic language models are not particularly useful for recognizing speech in any specific language domain. Typically, general use language models are improved for specific domains through the process of adaptation, in which a set of orthographies representative of the target domain are introduced to influence the various statistical weights in the model [6]. Adaptation has proven to be a successful method for increasing the accuracy of speech recognizers for specific domains, and remains a common approach for many ASR applications.

Practically speaking, individual voice command grammars are essentially a formatted language data set that covers a specific domain. The "domain" covered, in this case, simply consists of the contents of the grammar. Therefore, the resulting list of potential orthographies in a grammar can be used to adapt a general use language model to cover its specific "domain" more accurately. Additionally, since the data used for adaptation is identical to the expected speech input, the accuracy of in-grammar utterances will tend to be relatively high when using the adapted model. Probability scores are typically consistent for both in-grammar and OOG utterances, due to the model's wide coverage range.

However, the relatively large size of general use language models limits adaptation and training to desktop or server based hardware, and still takes a considerable amount of time to complete. These larger language models also require a sizeable amount of storage space and system memory to operate (>250 MB), and typically take upwards of 30–40 s to initiate on a mobile device. Once started, the response time from these larger models, especially for longer utterances, is considerably longer than is necessary to support an always listening voice command configuration (>2-3 s per utterance).

On the other end of the spectrum, generating an entirely new language model for each voice command grammar results in a much smaller and manageable language model. By using only the orthographies from the grammar to train the model, the size of the model is determined by the size of the grammar. Even the most complex grammars contain only a tiny fraction of the data that is contained a typical general use language model, resulting in a model that can be generated on a mobile device, and that responds quickly to each utterance. Again, since the training data matches the expected speech input, accuracy for in-grammar utterances is very high.

Unfortunately, the limited coverage of the language model, which allows for its high level of efficiency, adversely affects the recognizer's ability to deal with OOG utterances in a consistent matter. Since the model is based solely on orthographies from the grammar, the recognizer must assume that any utterance it hears must consist of words contained in the grammar. Since the vast majority of utterances that an always listening voice command system receives are OOG, the inevitable result is a high level of false-positive recognitions, along with wildly varying probability scores for each recognition result.

Based upon the known advantages and limitations of the two prior language model approaches, the SMASH team hypothesized that a balance could be struck between the two to achieve the desired behavior from the recognizer. By augmenting the training data obtained from the grammar with additional orthographies, the team predicted that it could introduce enough additional variation in to the resulting language model to lower the frequency of in-grammar false-positives to an acceptable level, while maintaining a high level of accuracy for in-grammar utterances.

The initial test of this hypothesis focused on generating a language model to recognize commands from a large grammar (>35000 unique orthographies, ~1500 unique words). Once the orthography list had been extracted from the grammar, using the SMASH GetOrthoList tool, an additional set of orthographies were added. The additional orthography list consisted of approximately 120,000 unique orthographies and 150 unique words. All of the words and orthographies in the combined set were unique in this case. The combined orthography list was then used to train a new language model, using the SMASH GenerateLM tool. The team inserted the resulting language model and dictionary, along with the original voice command grammar, into the SMASH Test Suite for evaluation.

Initial automated tests, which included providing pre-recorded in-grammar utterances, OOG utterances, and OOV utterances to the system, showed satisfactory results. The in-grammar utterances were recognized with consistently high probability scores, while all OOG and OOV utterances were handled without returning any in-grammar false positives. While performance at runtime was acceptable, on-device language model generation time was longer than expected, at over 15 s. The team then began iterative pruning and testing, reducing the size of the additional orthography list each time to reduce the time necessary for language model generation. When the additional orthography list was reduced to approximately 60,000 orthographies, the language model generation time had been reduced to around 5 s, and the recognition response time had been noticeably improved as well. Throughout the tests, recognition accuracy remained consistent during automated testing.

After this initial success, a remaining concern with this augmented orthography list approach was the applicability of the data set to different grammars. The team was also concerned about the impact of overlapping words and orthographies. Based on the initial results with a very large orthography list, and the steady level of accuracy as the list was shortened, the team had predicted that the same orthography list could provide similar results for small grammars. However, additional testing showed that using a fixed set of additional orthographies was ineffective for small grammars, reducing accuracy of in-grammar utterances to an unusable level. The accuracy degradation was especially worse for grammars that shared words and orthographies with the set of additional orthographies. These findings led to the determination that while augmenting a grammar-based language model with additional orthographies improved performance for always listening voice commands, using a fixed data set was not a viable implementation. In order to achieve the desired results, a unique set of orthographies would need to be generated for every grammar-based language model generated by SMASH.

Next, the SMASH team analyzed the relationship between the original grammar and orthography list used in the successful implementation described in the previous section. Based on the outcome of tests involving grammars of various sizes/contents, the team determined that there were three factors that impacted performance of the resulting language model. The first factor is the proportion of unique words between the voice command grammar and the additional orthography list. The second factor is the proportion of unique orthographies between the two sets. The third factor was the overlap of words and orthographies between the two sets. The team hypothesized that it could implement rules for generating an orthography list that would control these factors, keeping them consistent with those in the reference grammar and orthography list, which would result in a satisfactory language model.

The following rules, which define the values for each of the three factors, describe the algorithm that is implemented by SMASH in the creation of an automatically generated orthography list for any input grammar. These automatically generated orthography lists are referred to as Automated Model Optimization Data (AMOD).

- The AMOD shall contain one (1) unique word for every five (5) unique words contained in the input grammar, with a minimum value of five (5) unique words.
- The AMOD shall contain two (2) unique orthographies for every one (1) unique orthography contained in the input grammar, with a minimum value of 25 unique orthographies.
- The AMOD shall contain no words that are present in the input grammar.
- The AMOD shall contain an even distribution of 1-gram, 2-gram, and 3-gram orthographies.

The words used in the creation of the AMOD are contained in a list of randomly generated English words located inside the SMASH library. The list contains uncommon words, and in some cases, words which are pronounceable but are not found in any English dictionary. The list was generated using an open source web tool [7], and was then edited by the SMASH team to remove any potentially troublesome words. SMASH selects words from the list, and generates random orthographies based upon the aforementioned algorithm to produce the AMOD that will be used to augment the voice command grammar.

Before the language model is trained, the AMOD must be combined with the input grammar's orthography list. The grammar's orthography list is duplicated at this stage in order to add additional weight to the phrases contained in the grammar. This final combined list is used as training data for generating the language model that will be used by SMASH for recognizing the voice commands contained in the grammar.

The overall AMOD rule set has proven to be effective for use with voice command grammars of varying sizes and linguistic complexity by striking a balance between false-positive frequency, in-grammar utterance accuracy, and consistency of probability scores. However, the ratios used in the current SMASH AMOD algorithm may require further adjustment to optimize performance for those (or other) metrics and for specific data sets, and are generally a good starting point rather than a final set of values.

5 On-the-Fly Automated Language Model Creation in SMASH

SMASH uses a unique, original algorithm to generate language models and dictionaries, on the fly and on-device, based on an input SMASHBNF grammar. The generated language model is in the well-known ARPA-MIT LM format [8], and the generated dictionary is in the same format as the CMU Pronouncing dictionary [9]. The CMU PocketSphinx speech recognizer utilized in SMASH takes both of these files as input, and then decodes and recognizes speech based on the content of these two files. We will use the same simple grammar used in the previous example.

SMASH uses the following seven steps to create the language model and dictionary:

5.1 Generate Orthographies

To properly generate a language model, a reference text that contains all the possible utterances for a grammar must be created. As the first step in language model generation, SMASH creates this text. This text is called the orthography list, and is generated by recursively traversing a grammar starting with the start rule. For grammar rules that use the Kleene star (*) or plus operator (+), the utterance is included only once. For the sample grammar, here is the orthography list:

```
<s> MICROPHONE CHECK </s>
<s> MIKE CHECK </s>
<s> TESTING THE MICROPHONE </s>
<s> TESTING THE MIKE </s>
<s> TRYING OUT VOICE COMMANDS </s>
```

The <s> and </s> tags are used to delimit utterances in the orthography list. SMASH capitalizes all orthographies to perform case-specific comparisons against definitions in the CMU pronouncing dictionary.

5.2 Generate Vocabulary

A list of unique words used in the grammar is needed for language model generation. This is called the vocabulary of the grammar. The vocabulary can be deduced from the orthography list, by keeping a set of words for the grammar. The vocabulary for the sample grammar is:

```
</s>
<s>
CHECK
COMMANDS
MICROPHONE
MIKE
OUT
TESTING
THE
TRYING
VOICE
```

5.3 Pick and Determine Number of AMOD Words

The key to generating a useful language model is to have an AMOD word list that has words that are distinct from words in the grammar *(See section Automated Model Optimization Data (AMOD))*. SMASH reads in the list of AMOD words, and picks words in the order they were read in. SMASH uses a 1:5 ratio to determine the number of words to pick (with a minimum of 5). For example, if a grammar's vocabulary contains 100 words, 20 AMOD words will be picked from the AMOD list. If an AMOD words matches a word found in the grammar, that AMOD word is discarded.

In this case, five words were picked since that is the minimum number of AMOD words that we can use. Also, the AMOD words are added to the vocabulary list generated in Sect. 2.

5.4 Generate AMOD Orthographies

We will now generate orthographies, using the AMOD word data, which will be injected into the language model generation process. SMASH generates twice as many AMOD orthographies as there are in the grammar's orthography list (or a minimum of 25). This allows us to increase the variety and effectively reduce the weighting of the individual AMOD orthographies. The generation process evenly creates one-third 1-grams (one word orthographies), one-third 2-grams (two word orthographies), and one-third 3-grams (three word orthographies). All of the AMOD orthographies are randomly generated, which means that running the generation process multiple times will yield different results.

5.5 Create a Final Orthography List

We will now create a final integrated list of orthographies, which contain both the AMOD orthographies and the actual orthographies from the grammar. SMASH gives more weight to the actual orthographies, by using all of the actual orthographies twice when creating the final list. For example, in Sect. 1, we generated five orthographies. Therefore, SMASH will double these orthographies.

We will also add in all of the randomly generated AMOD orthographies that were generated in Sect. 4. This final list is what SMASH uses to generate the language model and dictionary used by the CMU PocketSphinx speech recognizer.

5.6 Generate the Language Model

SMASH uses two software tools from the CMU-Cambridge Language Modeling Toolkit (CMUCLMTK) to transform the final orthography list into a language model [10].

5.7 Generate the Dictionary

The dictionary file provides a list of pronunciations (word to phoneme mappings) for every word that appears in the language model. The dictionary file consists of one word per line, along with its ARPAbet pronunciation. SMASH generates a dictionary file from three sources: (1) custom pronunciations in the grammar file, (2) the CMU Pronouncing Dictionary, and (3) using letter-to-sound rules. The vocabulary file generated in Sect. 2 is used as input.

6 Future Work on SMASH and AMOD

With the implementation of the AMOD in the SMASH workflow, the SMASH team was able to realize comparable levels of voice command accuracy and efficiency to that of commercial products using widely available open-source tools. However, there are gains in performance and efficiency yet to be made. The SMASH team plans to continue refining the AMOD process through the addition of phonemic analysis, and through continued testing and integration to optimize the existing AMOD algorithm. This future work will continue to build on the current improvements and focus on maximizing the benefit of open-source speech and language tools.

References

1. Intent | Android Developers. https://developer.android.com/reference/android/content/intent.html
2. JSpeech Grammar Format. https://www.w3.org/TR/jsgf/
3. CMU Sphinx. http://cmusphinx.sourceforge.net
4. Services | Android Developers. https://developer.android.com/guide/components/services.html
5. JSGFGrammar (Sphinx-4). http://cmusphinx.sourceforge.net/doc/sphinx4/edu/cmu/sphinx/jsgf/JSGFGrammar.html
6. Bellegarda, J.R.: An overview of statistical language model adaptation. In: ITRW on Adaptation Methods for Speech Recognition. Sophia Antipolis, France (2001)
7. Random Word Generator. https://randomwordgenerator.com
8. The ARPA-MIT LM format. http://www1.icsi.berkeley.edu/Speech/docs/HTKBook3.2/node213_mn.html
9. The CMU Pronouncing Dictionary. http://www.speech.cs.cmu.edu/cgi-bin/cmudict
10. Clarkson, P.: The CMU-Cambridge Statistical Language Modeling Toolkit v2. http://www.speech.cs.cmu.edu/SLM/toolkit_documentation.html
11. FreeTTS Programmer's Guide. http://freetts.sourceforge.net/docs/ProgrammerGuide.html
12. Black, A.W., Lenzo, K., Pagel, V.: Issues in building general letter-to-sound rules. In: Proceedings of ECSA Workshop on Speech Synthesis, pp. 77–80, Australia (1998)

The Modeling of Technological Trade-off in Battery System Design Based on an Ergonomic and Low-Cost Alternative Battery Technology

Victor K.Y. Chan(✉)

School of Business, Macao Polytechnic Institute,
Rua de Luis Gonzaga Gomes, Macao SAR, China
vkychan@ipm.edu.mo

Abstract. The author and a co-inventor published earlier an ergonomic and low-cost alternative battery technology in a bid to tackle batteries' long-standing conundrums of slow charging and costly charging infrastructures. The technology adopts modularization of battery systems and prioritization of the charging/discharging of battery module(s) according to users' human effort to unload/(re)load individual battery module(s), enabling users to unload preferentially, substantially discharged battery module(s) individually for (re)charging them with the domestic mains supply before reloading them instead of (re)charging the whole battery system in situ. As the charging speed is irrelevant to domestic charging, for example, at home overnight, which also avoids specialized charging infrastructures, the two conundrums above can no longer be concerns. However, this technology is confronted with a trade-off of users' human effort between unloading/reloading many small battery modules and unloading/reloading fewer larger and heavier battery modules. This paper models this trade-off for the ultimate goal of optimization.

Keywords: Battery systems · Trade-off · Human factors · Ergonomics

1 Introduction

For decades, batteries have been used for powering electrical and electronic equipment, inclusive of instruments, devices, and machineries. Batteries' role in powering mobile phones, tablet computers, other mobile information technology devices, and electrical vehicles is especially in the limelight these days. All such equipment comprises and/or is connected to housings where batteries are usually installed and is thus electrically powered by the batteries' stored energy released during their discharge. Conversely, the stored energy in the batteries is replenished through (re)charging them. In view of its much lower hazardous emission level and operational cost, battery power is far more environmentally friendly [1, 2] and economical [2] than most other means of mobile power.

Ever since batteries' advent, scientists in academia and industry have been researching into material science, battery microstructures, and battery manufacturing

T. Ahram and W. Karwowski (eds.), *Advances in Human Factors, Software, and Systems Engineering*, Advances in Intelligent Systems and Computing 598,
DOI 10.1007/978-3-319-60011-6_12

technologies to ameliorate battery chemistry, anode/cathode materials, battery manufacture optimization [3] so as to give rise to a desirable trade-off between such major properties of batteries as

(a) the cycle life (i.e., the number of charge/discharge cycles to undergo before the energy capacity degrades significantly) [4],
(b) the power density (i.e., the maximum amount of power supplied per unit volume) [5],
(c) the energy density (i.e., the energy capacity per unit volume or mass) [6],
(d) the capital cost per unit energy capacity [7],
(e) the charging speed (or, equivalently, the time to (re)charge the batteries from a lower state of charge to a higher state of charge) [8], and
(f) the capital cost of charging infrastructures [9].

In particular, battery applications are especially beset by the limitations on the last two properties (e) and (f) above [10].

2 The Ergonomic and Low-Cost Alternative Battery Technology

The author together with a co-inventor published earlier an ergonomic and low-cost, alternative technology in a bid to practically and industrially eradicate the two aforementioned limitations without recourse to any revolutionary technological breakthroughs in material science, battery microstructures, or battery manufacturing technologies [11]. Leaving aside its various collateral concepts, the technology's key idea is to adopt modularization of battery systems, prioritization of the discharge of battery modules in a battery system according to users' human effort to unload the battery modules from and/or to (re)load the battery modules into the housing and/or to load replacement battery modules into the housing such that discharge priority is accorded to battery modules in a battery system in descending order of their ease of being unloaded and/or (re)loaded and/or the ease of their replacement battery modules being loaded when the battery modules are discharged to power electrical or electronic equipment. Consequently, by the time the battery system needs (re)charging, only a subset of such preferentially discharged battery modules, as opposed to the battery system as a whole, are substantially discharged and in need of (re)charging. Thus, users may opt to individually unload from the housing only such preferentially, substantially discharged battery modules for conveying them to and (re)charging them with, for example, the domestic mains supply before individually (re)loading them back into the housing instead of simply (re)charging the whole battery system in situ. Alternatively, the users may opt to individually unload from the housing only such preferentially, substantially discharged battery modules and individually load fully charged replacement battery modules into the housing instead of unloading the battery system as a whole and loading a replacement battery system as a whole into the housing. As the charging speed is rather irrelevant to domestic charging, for example, at the users' homes overnight, which also obviates the need for specialized charging infrastructures, limitations on (e) and (f) above can no longer be concerns.

[1, 1]	[1, 2]	…	[1, q - 1]	[1, q]
[2, 1]	[2, 2]	…	[2, q - 1]	[2, q]
…	…	…	…	…
[p - 1, 1]	[p - 1, 2]	…	[p - 1, q - 1]	[p - 1, q]
[p, 1]	[p, 2]	…	[p, q - 1]	[p, q]

Fig. 1. This is an exemplary layout of the battery modules in a modularized battery system, which comprises $p \times q$ battery modules. Each cell denotes a battery module whereas the thick outer frame denotes the housing. Without loss of generosity, assume that the housing's opening is at the top. Also, despite the columns appearing vertical, they can be alternatively oriented, for example, horizontal.

It is noteworthy that it is the modularization of the battery system, for example, as in Fig. 1 that in essence breaks the otherwise much larger battery system into smaller battery modules, likely ending up with the relatively small battery modules being loadable and unloadable individually even in the absence of specialized loading/unloading facilities. Likewise, thanks to modularization, domestic charging of individual battery modules becomes likely plausible whilst charging the otherwise battery system as a whole certainly calls for specialized charging facilities of a much higher power and thus much more sophisticated safety protection. In addition, prioritization of the discharge of battery modules renders the subset of preferentially, substantially discharged battery modules in need of (re)charging likely to be those easily unloadable and/or (re)loadable and/or those of which replacement batteries are easily loadable, further relieving the users' human effort to unload and (re)load such battery modules for (re)charging or to load replacement batteries and further obviating the need for specialized loading/unloading facilities.

Notwithstanding its aforesaid advantages, this technology is confronted with an unprecedented trade-off of users' human effort between unloading/conveying/(re)loading a large number of battery modules of small masses/sizes and unloading/conveying/(re)loading a small number of battery modules of large masses/sizes. In an attempt to further this technology's development, this article devises a mathematical model of such a trade-off by assuming a Poisson probability distribution of the number of substantially (and preferentially) discharged battery modules in need of (re)charging at the end of a charge cycle of a battery system, an arithmetic progression-based relationship between the number of such substantially discharged battery modules and the human effort to unload and (re)load them, and a power relationship between the mass of each such battery module and the human effort to convey it.

3 The Trade-off Model

Modeling by the Poisson distribution [12, 13] the probability distribution of the number of substantially discharged battery modules in need of (re)charging at the end of a charge cycle of a battery system (among a batch of battery systems), i.e., the probability distribution of the number of substantially discharged battery modules in need of (re) charging immediately before a charging exercise of a battery system among that batch of battery systems, the corresponding probability distribution function is given by

$$P(K = k) = \frac{\lambda^k e^{-\lambda}}{k!} \tag{1}$$

where

K = the number of substantially discharged battery modules in need of (re) charging at the end of a charge cycle of a battery system,

$P(K = k)$ = the probability of the event that the number of substantially discharged battery modules in need of (re)charging at the end of a charge cycle of a battery system is equal to k, and

λ = the *mean* number of substantially discharged battery modules in need of (re)charging at the end of a charge cycle of a battery system.

In a particular scenario being studied, the population is all the charge cycles of all the battery systems among a particular batch of battery systems and is characterized by the parameter λ, which is in turn dependent on the energy consumption of the equipment powered by the battery systems, the equipment usage and charging habits of the users, and the energy capacities of the battery modules (which are in turn dictated by the type of the battery modules and their degrees of aging). Therefore, λ in a particular scenario can only be determined empirically as estimated by a sample mean number of substantially discharged battery modules in need of (re)charging at the end of a charge cycle of a battery system. It is noteworthy that a real-life scenario being studied is typified by an investigation into a batch of battery systems powering some equipment units of a particular brand and model, so all the battery modules and all the battery systems are presumably and typically identical (probably, except for their levels of aging) as assumed throughout the rest of this article. This way, the aforementioned sample mean can be calculated by summing the numbers of substantially discharged battery modules in need of (re)charging at the end of a sample of charge cycles from that batch of battery systems during the normal operation of the equipment being powered and dividing the sum by the number of such charge cycles experienced.

In view of battery modules being usually installed in their battery system's housing in a layered layout as in Fig. 1, the battery modules in the layer least deep inside the housing (i.e., those closest to the housing's opening or [1, 1] to [1, q]) are the easiest to unloaded from and (re)loaded into the housing, those in the layer second least deep inside the housing (i.e., [2, 1] to [2, q]) are the second easiest to unloaded and (re) loaded, and so on whilst the battery modules in the layer deepest inside the housing (i.e., those farthest from the housing's opening or [p, 1] to [p, q]) are the least easy to unloaded. Hence, discharge priority is also accorded in the above order with [1, 1] to [1, q] having the highest priority, [2, 1] to [2, q] the second highest, and [p, 1] to [p, q] the lowest, and thus the battery modules during the normal operation of the equipment become substantially discharged and in need of (re)charging also in the above order with [1, 1] to [1, q] becoming so first, [2, 1] to [2, q] next, and [p, 1] to [p, q] last. Reasonably assuming that the difference in the ease of unloading and (re) loading battery modules between those in any two consecutive layers is a constant, users' human effort to unload and (re)load k substantially discharged battery modules in need of (re)charging at the end of a charge cycle of a battery system is given by

$$E_{\mathrm{L}} = E_{\mathrm{L1}} + (E_{\mathrm{L1}} + E_{\mathrm{LD}}) + (E_{\mathrm{L1}} + 2E_{\mathrm{LD}}) + (E_{\mathrm{L1}} + 3E_{\mathrm{LD}}) + \cdots + [E_{\mathrm{L1}} + (l-1)E_{\mathrm{LD}}] + \frac{r}{L}(E_{\mathrm{L1}} + lE_{\mathrm{LD}}) \tag{2}$$

$$= \frac{l}{2}[2E_{\mathrm{L1}} + (l-1)E_{\mathrm{LD}}] + \frac{r}{L}(E_{\mathrm{L1}} + lE_{\mathrm{LD}}) \tag{3}$$

$$= \frac{\left\lfloor \frac{k}{L} \right\rfloor}{2}[2E_{\mathrm{L1}} + (\left\lfloor \frac{k}{L} \right\rfloor - 1)E_{\mathrm{LD}}] + \frac{\left\lceil \frac{k}{L} \right\rceil}{L}(E_{\mathrm{L1}} + \left\lfloor \frac{k}{L} \right\rfloor E_{\mathrm{LD}}) \tag{4}$$

where

E_{L} = users' human effort to unload and (re)load $k = lL + r$ battery modules,

l = the (integral) number of layers fully occupied by the k battery modules,

L = the number of battery modules in a layer,

r = the (integral) number of battery modules in the layer partially occupied by the k battery modules such that $0 \leq r < L$,

E_{L1} = users' human effort to unload and (re)load all the battery modules in the layer with the most easily unloadable and (re)loadable battery modules, i.e., in the layer least deep inside the housing or the layer nearest to the housing's opening, and

E_{LD} = the (incremental) difference in users' human effort to unload and (re)load all the battery modules in a layer between any two consecutive layers.

$\left\lfloor \frac{k}{L} \right\rfloor$ = the quotient of k divided by $L = l$, and

$\left\lceil \frac{k}{L} \right\rceil$ = the remainder of k divided by $L = r$.

Please note that the equality between (2) and (3) is due to all the terms in (2) other than the last one being an arithmetic progression. L is known to the users given the layout of the battery modules in the battery system. E_{L1} and E_{LD} are determined by users' corresponding ratings or as the means of a sample of users' corresponding ratings.

It is conceivable that users' human effort to convey a battery module, for example, when being unloaded from the housing, (re)loaded into the housing, or carried between the housing and the charging facility, increases only marginally even if its mass more than doubles when its mass remains on the low side but more than doubles even if its mass increases only marginally when its mass is approaching the physical limit of the users. Hence, users' such human effort to convey a battery module typically relates to its mass through a power function with a power greater than one as follows:

$$E_{\mathrm{C}} = m^{s} \tag{5}$$

where

E_C = users' human effort to convey the battery module,

m = the mass of the battery module,

s = the power of the power function, which is greater than 1 to model the aforesaid pattern of the increase in users' human effort with the increase in the battery module's mass.

Please note that the word "power" in this paragraph refers to its mathematical meaning and has nothing to do with electrical power of batteries or otherwise. In practice, m is known to the users upon weighing, and s can be estimated as the regression coefficient in the regression equation

$$\ln E_C = s \ln m \tag{6}$$

using multiple value pairs of (E_C, m), or equivalently ($\ln E_C$, $\ln m$), which are in turn obtained from users' different ratings of the human effort E_C to convey battery modules of different masses m.

Adhering to the assumption of identical battery modules (probably, except for their levels of aging) in identical battery systems (probably, except for their levels of aging) in question, all things being equal, if the mass m of each battery module in the battery systems doubles, then the energy capacity of each battery module doubles, half the number k of battery modules are needed to be substantially discharged to power equipment with the same energy consumption, and thus the *mean* number λ of substantially discharged battery modules in need of (re)charging at the end of a charge cycle of a battery system halves. As such, m is inversely proportional to λ, all things being equal, or mathematically,

$$m\lambda = C \tag{7}$$

where C is a constant, which can be determined by multiplying a fixed, known m value by the λ value empirically estimated for (1) when the mass of each battery module is equal to that m value. Alternatively, and more precisely, C can be estimated as the mean of the multiple values of C, each determined through the multiplication above but with a different fixed, known m value and thus a different λ value empirically estimated.

Consequently, (1) becomes

$$P(K = k) = \frac{\left(\frac{C}{m}\right)^k e^{-\frac{C}{m}}}{k!} \tag{8}$$

Therefore, the probabilistic expected value of users' total human effort to spent on (re)charging a battery system at the end of any of its charge cycles is given by

$$
\begin{aligned}
E &= E(E_{\mathrm{L}} + kE_{\mathrm{C}}) \\
&= \sum_{k=0}^{n} \left\{ \begin{array}{c} \dfrac{\left\lfloor \frac{k}{L} \right\rfloor}{2}\left[2E_{\mathrm{L1}} + \left(\left\lfloor \dfrac{k}{L} \right\rfloor - 1\right)E_{\mathrm{LD}}\right] \\ + \dfrac{\left\lceil \frac{k}{L} \right\rceil}{L}\left(E_{\mathrm{L1}} + \left\lfloor \dfrac{k}{L} \right\rfloor E_{\mathrm{LD}}\right) + km^{s} \end{array} \right\} P(K = k) \\
&= \sum_{k=0}^{n} \left\{ \begin{array}{c} \dfrac{\left\lfloor \frac{k}{L} \right\rfloor}{2}\left[2E_{\mathrm{L1}} + \left(\left\lfloor \dfrac{k}{L} \right\rfloor - 1\right)E_{\mathrm{LD}}\right] \\ + \dfrac{\left\lceil \frac{k}{L} \right\rceil}{L}\left(E_{\mathrm{L1}} + \left\lfloor \dfrac{k}{L} \right\rfloor E_{\mathrm{LD}}\right) + km^{s} \end{array} \right\} \dfrac{\left(\frac{C}{m}\right)^{k} e^{-\frac{C}{m}}}{k!}
\end{aligned} \tag{9}
$$

where

E = the expected value of users' total human effort to (re)charge a battery system at the end of any of its charge cycles,

$E(\cdot)$ = the probabilistic expected value function, and

n = the total number of battery modules in each battery system.

In practice, n is known to the users given the layout of the battery modules in each battery system.

4 Ultimate Objective: Optimization

The ultimate objective of this modeling is of course to optimize, or more precisely minimize, E in (9) upon substituting into it the values of n, L, E_{L1}, E_{LD}, s, and C as found in ways delineated in the last section presumably and typically for battery systems powering equipment units of a particular brand and model. Mathematically, the minimization is

$$
\min E = \min_{m \in \left\{M_{\mathrm{L}}, \frac{M_{\mathrm{L}}}{2}, \frac{M_{\mathrm{L}}}{3}, \frac{M_{\mathrm{L}}}{4}, \cdots\right\}} \sum_{k=0}^{n} \left\{ \begin{array}{c} \dfrac{\left\lfloor \frac{k}{L} \right\rfloor}{2}\left[2E_{\mathrm{L1}} + \left(\left\lfloor \dfrac{k}{L} \right\rfloor - 1\right)E_{\mathrm{LD}}\right] \\ + \dfrac{\left\lceil \frac{k}{L} \right\rceil}{L}\left(E_{\mathrm{L1}} + \left\lfloor \dfrac{k}{L} \right\rfloor E_{\mathrm{LD}}\right) \\ + km^{s} \end{array} \right\} \dfrac{\left(\frac{C}{m}\right)^{k} e^{-\frac{C}{m}}}{k!} \tag{10}
$$

where M_{L} = the total mass of each layer of battery modules in each battery system, and is accomplished by searching for the optimal m out of the feasible values M_{L}, $\frac{M_{\mathrm{L}}}{2}$, $\frac{M_{\mathrm{L}}}{3}$, $\frac{M_{\mathrm{L}}}{4}$, …, which correspond to splitting each layer into different (integral) numbers of battery modules. In practice, this search needs not to be indefinite but can stop at the practical minimum mass of each battery module as limited by its manufacture and daily usage.

The search can be performed simply by substituting M_{L}, $\frac{M_{\mathrm{L}}}{2}$, $\frac{M_{\mathrm{L}}}{3}$, $\frac{M_{\mathrm{L}}}{4}$, … into m to find the corresponding minimum E value or by means of genetic algorithms.

5 Discussion

This article aims to propose (9) as a mathematical model of technological trade-off in battery system design based on an ergonomic and low-cost alternative battery technology that was published earlier by the author and a co-inventor and is now covered by several patents pending around the globe. It also suggests (10) for the optimization, or more precisely, minimization of users' human effort to spent on (re)charging the battery systems in question in the context of the aforementioned trade-off through the search for the corresponding optimal mass of each battery module. This way, the article serves to further the technology's ideas by pondering one top-priority aspect for the technology's real-life manufacture and commercialization in the highly competitive market, i.e., the optimization of one of its key ideas' parameters concerning the mass m of each battery module in the battery systems upon the modularization of the battery systems for the sake of users' best ergonomic benefit in spending the minimum human effort to (re)charge the battery modules at the end of each charge cycle.

Whereas the author is still in the process to try to empirically perform the optimization, such empirical work must be specific to the battery systems of the equipment units of a particular brand and model, so large-scale cooperation with the industry is essential and is what the author is still striving for. Once this is accomplished, the results will very likely appear in the author's further publications.

Acknowledgments. This article was supported by Grant (004/2013/A) of the Science and Technology Development Fund, the Government of Macao Special Administrative Region.

References

1. Hülsmann, M., Fornahl, D. (eds.): Evolutionary Paths Towards the Mobility Patterns of the Future. Lecture Notes in Mobility. Springer, Berlin (2014). doi:10.1007/978-3-642-37558-3
2. Shiau, C.-S.N., Samaras, C., Hauffe, R., Michalek, J.J.: Impact of battery weight and charging patterns on the economic and environmental benefits of plug-in hybrid vehicles. Energy Policy **37**, 2653–2663 (2009)
3. Lin, M.-C., Gong, M., Lu, B., Wu, Y., Wang, D.-Y., Guan, M., Angell, M., Chen, C., Yang, J., Hwang, B.-J., Dai, H.: An ultrafast rechargeable aluminium-ion battery. Nature **520**, 324–328 (2015)
4. Omar, N., Monem, M.A., Firouz, Y., Salminen, J., Smekens, J., Hegazy, O., Gaulous, H., Mulder, G., Van den Bossche, P., Coosemans, T.: Lithium iron phosphate based battery – assessment of the aging parameters and development of cycle life model. Appl. Energy **113**, 1575–1585 (2014)
5. Shousha, M.: Design and implementation of high power density assisting step-up converter with integrated battery balancing feature. IEEE J. Emerg. Sel. Top. Power Electron. **PP**, 1 (2017)
6. Liu, Q.-C., Liu, T., Liu, D.-P., Li, Z.-J., Zhang, X.-B., Zhang, Y.: A flexible and wearable lithium–oxygen battery with record energy density achieved by the interlaced architecture inspired by bamboo slips. Adv. Mater. **28**, 8413–8418 (2016)
7. Nykvist, B., Nilsson, M.: Rapidly falling costs of battery packs for electric vehicles. Nat. Clim. Change **5**, 329–332 (2015)

8. Hsieh, G.-C., Chen, L.-R., Huang, K.-S.: Fuzzy-controlled Li-ion battery charge system with active state-of-charge controller. IEEE Trans. Ind. Electron. **48**, 585–593 (2001)
9. Yilmaz, M., Krein, P.T.: Review of battery charger topologies, charging power levels, and infrastructure for plug-in electric and hybrid vehicles. IEEE Trans. Power Electron. **28**, 2151–2169 (2013)
10. Ralston, M., Nigro, N.: Plug-in electric vehicles: literature review. Technical report, Center for Climate and Energy Solutions (2011)
11. Chan, V.K.Y., Leong, S.L.: Novel, low-cost alternative technologies to tackle practical, industrial conundrums – a case study of batteries. MATEC Web of Conf. **44** (2016). 02096
12. Poisson, S.D.: Recherches sur la Probabilité des Jugements en Matière Criminelle et en Matière Civile, Précédées des Règles Générales du Calcul des Probabilitiés. Bachelier, Paris, France (1837)
13. Render, B., Stair Jr., R.M., Hanna, M.E., Hale, T.S.: Quantitative Analysis for Management, 12th edn. Pearson, Essex (2015)

The Construction of the Picture Database for Different Ethnic Life Styles: A Case Study of Lifestyles of Health and Sustainability (LOHAS)

Peng-Jyun Liu[1,3(✉)], Ching-yi Wang[2], and Ming-Chuen Chuang[3]

[1] Department of Creative Product Design, Asia University, 500, Lioufeng Rd., Wufeng, Taichung 41354, Taiwan, ROC
a804338@gmail.com

[2] Department of Industrial Design, Tatung University, No. 40, Sec. 3, Zhongshan N. Rd., Taipei City 104, Taiwan, ROC

[3] Institute of Applied Arts, National Chiao Tung University, 1001 University Road, Hsinchu 300, Taiwan, ROC

Abstract. Emotional attraction has been a major motive for consumers to purchase commodities. As the rising of consumers' purchase motive is one of the main design objectives, affective attraction is the key core of product design. To realize the objective, designers have adopted many methods in the design process. Among these methods, it is the most common one to stimulate ideas with picture boards which can be used to directly express and distinguish product pictures and also as a tool of affective evaluation. The research targets to the life style boards proposed by Baxter, as they are widely used. The research aims to build a database which can provide proper pictures to express different life style ethnics' life images, waste a huge amount of time and reduce the burden of artificially collecting pictures. In the research, 100 consumers are firstly invited for screening the life style ethnics and 23 representatives selected. Then, 6 experts and scholars from different fields are invited to select the adjective words and pictures for the life style ethnics, and then the SD evaluation investigation is made among the 23 representatives to obtain the adjective words and pictures of the LOHAS' life styles. The research result will significantly influence the design process, improve the product design efficiency, increase the expression accuracy of design pictures and allow more people to participate in the design process of image boards. Besides, the research result can be extended to other fields, including industry, manufacturing industry or education sector, which can increase users' happiness and satisfaction, and improve the product market competitiveness.

Keywords: Picture database · Image evaluation · SD (Semantic difference Method) · LOHAS

1 Introduction

It is an indispensable part of design activities to analyze users' expectations and demands, and direct immersion in users' feelings can help designers the design problems and avoid conventionalized design [1] Many researches point out that

T. Ahram and W. Karwowski (eds.), *Advances in Human Factors, Software, and Systems Engineering*, Advances in Intelligent Systems and Computing 598,
DOI 10.1007/978-3-319-60011-6_13

consumers can obtain more pleasure and satisfaction by buying the products with the image board design [2]. To achieve the objective, besides importing the image boards in design process at the design conception stage, image boards can be also applied to achieve the product differentiation and shorten the product development cycle in the mature period of product design.

To achieve the objective, the designers have adopted many methods in the design process. Among it is the most common one to stimulate ideas with picture boards. Picture boards can be used to directly express and distinguish product pictures and also as a tool of affective evaluation. Therefore, it is of positive significance for the entire product design to import the image board method in the design procedure [3]. The design method of using image boards in the design procedure can obtain users' sentimental demands and expected use benefits, and make product design satisfy the use groups' psychological or emotional demands.

2 Literature Review

At present, most of the research results on the image board are based on the arguments proposed by Baxter in Product Design and Development (1995) [4] which divides image boards into life-style board, mood board and theme board which have the progressive relations. According to Baxter, product design is an activity after visual thinking, dominated by visual sense, and emphasizes that designers can make design products closer to user demands by following life-style board → mood board → theme board.

2.1 Image Board

Baxter proposed three categories in image board in 1995 and defined the three kinds of image boards. It proposed life-style board, mood board and theme board which were progressively relevant. Life-style board collects the target consumers' life image information and pictures (as shown in Fig. 1); mood board is from life-style board to define a single value of products (as shown in Fig. 2); and theme board is derived from mood board, and it is recognized by the designers that picture information can transmit the final images of product targets (as shown in Fig. 3).

Fig. 1. (Baxter 1995)

Fig. 2. (Baxter 1995)

Fig. 3. (Baxter 1995)

2.2 Image Map

Image map is the color image system proposed by the Japanese color research expert Shigenobu Kobayashi in his book of Color Marketing Strategies (1986) [5]. The system sets forth and master the different color properties with quantity (the distance between two mutual positions in the map) (as shown in Fig. 4). The entire basic framework is a cross-shaped position axis in the X-Y direction, and composed by two groups of opposite adjectives which are respectively allocated at both ends of X Axis and Y Axis. The operation procedures contain sample collection, image system paper preparation (6-section quadrille paper at both the vertical and horizontal directions, the central point is 0, 0, both ends of X Axis is cool/warm color, and those of Y Axis are soft and hard colors), and the cool/warm and strong/soft degree positioning and image similarity classification of various color samples.

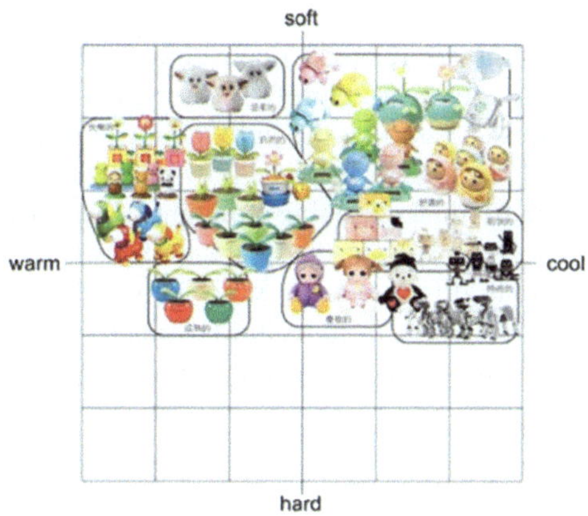

Fig. 4. Toy design image map (YI-RU LAN, 2006) [6]

The image boards have the following overall advantages, for example, they stimulate diversified creativity and product possibilities, enjoy superb communication and coordination effects, and can reach a consensus among designers, owners, personnel without the design background or design team members, and also revising and adjusting the design information sources in the product development process, etc.; the image map has the following advantages, for example, it has coordinate axes, the pictures are positioned precisely in the board making process, bases are available for picture grouping, the obtainable images are highly accurate, so it is greatly helpful to designers. Besides, it can be repeatedly checked and compared whether the picture positions are consistent with the sensibility, the objectiveness is superb, and the combined groups can be renamed and flexible; and speaking of its disadvantages, the positioning discussion is time-consuming, team members easily have disagreements in the picture cognition process, if disagreements occur, the pictures are positioned differently, and the picture position points can be only obtained in the form of mean value.

2.3 Images and Life Style

Images are generated in two ways, in the first way, they are directly obtained from the culture, including life experiences, social conventional norms, life backgrounds, etc.; and in the second way, they are the cognition which is conveyed to the brain indirectly through sensory stimuli, such as the imaginations after reading articles or thoughts after appreciating artistic works. Therefore, in a manner of speaking, objectively, images are the appearance data according to the sensory source, the processing images about cognitive objectives formed in the thinking space and the physical memory traces and overall structural relationships after the cognitive subjects come into contact with objective matters, and can be composed of internal sense and external form [7]. The majority of scholars in different fields think that images are composed cognition and affections [8], and also contain cognitive and emotional images. Cognitive images are about individual attitudes and impressions; and emotional images are about individuals' cognition of image attributes and features [9, 10]. The Japanese Kansei engineering expert Nagamachi (1995) [11, 12] also proposed that images are abstract but not concrete, and the associations are related to consumers' life experiences and cultural backgrounds, in other words, they are closely connected with personal life styles.

Life styles are theoretically based on the Individual Cognitive Structure Theory put forward by Kelly in 1955 [13] which explained how people organized their ego worlds inside according to their individual cognitive structures. Mr. Lazer (1963) [14] defined life styles as: [series concepts, and are the characteristics of the society or a certain social group in life which are different from those with other groups and concrete life behaviors. So life is the outcome from culture, values, resources, laws and other strength, the consumers' purchase and consumption behaviors mirror the social life styles].

2.4 Semantic Differential Scale

Osgood (1957) [15] presented Semantic Differential Scale in the product image research, and the method investigates testees' subjective evaluation, translate the stimuli's feelings into quantitative form to obtain the product images. Mr. Osgood divided product images into three attributes of self-evidence, structuredness and significance.

Semanteme is divided by virtue of the semantic space of the related adjectives to evaluate connotations of individuals' affections about concepts. The method was co-developed by Mr. Osgood (1957) and Mr. Hofstätter (1958) [16], and divided the concept into levels (are normally in seven levels but flexible). Then the affective judgment is discovered by analyzing factors. The evaluation measuring scale is the two-pole way, namely, in the positive and negative poles, several persons' evaluations of a concept are collected to determine the mean value. It is affirmed that there are three concepts used by Mr. Hofstätter to differentiate the personality factors, namely, positive extroversion, negative extroversion, and introversion. Osgood simplified the measuring scale into three aspects: (1) evaluation (good - bad); (2) potency (strong - weak); (3) activity (fast - slow). There are 76 matching adjectives in the standard factors.

Table 1. Standard factors for Osgood's semantic item

Evaluation		Potency		Activity	
Good	Bad	Strong	Weak	Active	Passive
Happy	Unhappy	Soft	Hard	Tense	Loose
Positive	Negative	Heavy	Light	Impassioned	Calm
Optimistic	Pessimistic	Thick	Thin	Noisy	Quite
Bright	Shady	Robust	Exquisite	Fast	Slow
Special	General	Big	Small	Hot	Cold
Excellent	Adverse	Masculine	Feminine	Sharp	Dull
Elegant	Vulgar	Dignified	Merciful	Excited	Tranquil

Houng thought that the Chinese adjectives had similar semantic structure factors (Table 1) [17]. The adjectives adopted in the research are generally known by people, and two pairs of adjectives are available for each factor.

Concept evaluators can decide the measuring scales of a concept depending on their individual subjectivity. For example, the word of [school life] can be evaluated by the following three pairs of adjectives, and their attitudes towards school life can be known from the rated grade, and the corresponding relation between the design image vocabulary and the life style ethnics can be obtained by apply Osgood's three measuring scale aspects.

Consumer preferences from different life style ethnics are researched, according to the life style ethnics or semantic vocabulary classifications, pictures not only improve the design efficiency but also save a lot of manpower and time, but also make designers and users communicate with each other without obstacle, and also improve the precision of image reference when users and designers participate in design together. Given this, the research defines the traits of the life style ethnics explicitly, and build the image board picture database which has the similar attributes.

3 Research Methods

The entire research planning is conducted in three stages. The first stage is to seek for the LOHAS representative. In the research, 100 consumers are invited and screened for the life style ethnics, in order to seek for the testee representatives for the second stage. The questionnaire design refers to the life style trait problems provided by E-ICP (Eastern Integrated Consumer Profile) [18] co-founded by Eastern Online and College of Commerce, National Chengchi University. Regarding the questionnaire implementation, in consideration of convenient sampling and the dual-property relations between designers and consumers, we cooperate with a university which sets up a design college in Taichung City and conduct the random paper investigation among the junior students above in the departments of the design college.

The second stage is about the planning of testee samples, evaluation measuring scales, etc. Six experts and scholars from different fields are invited to select the adjectives and pictures for the LOHAS experiment. The research team firstly select lots of adjectives about the LOHAS from related newspapers, journals, network data, survey research reports, etc., and make the word cards (9 cm × 5.4 cm) and picture cards (4 × 6-inch, 300dpi, and with the assistance of computer projection when necessary), and then experts screen from them. There is no limit for the quantity of adjectives, and the experts adopt the most matching adjectives after sufficient discussion; in consideration of the quantity of the questions in the questionnaire, it is regulated that experts can select 20 pictures at the most, and decide to use one picture or not after sufficient discussion.

The third stage is about investigations on the LOHAS and image word, and the LOHAS and picture cognition. Firstly, questionnaire design is conducted, and the questionnaire contents are divided into two parts, the first part is about the basic data, such as genders, ages, design specialties, educational degrees, etc., and the second part is about the investigation contents, including pictures, adjectives, evaluation measuring scales, etc., and pictures and adjectives are the results screened by the experts in the second phase, and evaluation, potency and activity among the standard factors of Osgood's semantic items are selected as the measuring scales, and the experts select two pairs of adjectives in various standard factors as the evaluation measuring scales, and Likert scale is adopted for conducting evaluation and measurement, besides, considering that scale digits induce the testees to have the concept of being positive or negative, so the testees are only asked to click which adjectives they are inclined to in the blank form. The investigation is also conducted in two intervals, the first one is the adjectives and evaluation measuring scales (as shown in Table 2), and the second one is the pictures and evaluation measuring scales (as shown in Table 3). The LOHAS representatives who are screened out in the first stage conduct the questionnaires online.

Table 2. Life style ethnic groups and image word investigation

Q1 : Please fill in your feelings about the adjective [pragmatic]	Optimistic □□□□□ Pessimistic Bright □□□□□ Shady Masculine □□□□□ Feminine Exquisite □□□□□ Robust Active □□□□□ Passive Noisy □□□□□ Quiet

Table 3. The LOHAS and picture cognition investigation (20 pictures are available for each ethnic group)

Q1 : Look at the picture below, and click the adjectives which can express your feelings 	Optimistic □□□□□ Pessimistic Bright □□□□□ Shady Masculine □□□□□ Feminine Exquisite □□□□□ Robust Active □□□□□ Passive Noisy □□□□□ Quiet

4 Research Results

To screen out the LOHAS representatives, 100 testees are invited to make questionnaires. Women and student testees aged from 19 to 24 years old are in the majority might because of cooperation with the University, all of them are literate, and most of them have the senior college/bachelor degree (as shown in Table 4), Question 1 to Question 5 in the questionnaire are about the LOHAS traits, among the 31 testees scored above 4, 23 ones agree to conduct the next stage, and one woman testee fails to complete the section of picture investigation, therefore, there are 22 picture respondents.

Table 4. Percentages of basic questions and answers

Question	Option	Percent	Question	Option	Percent
1. Gender	Male	39%	3. Work Status	Students	92%
	Female	61%		Office Workers	4%
2. Age	19 -24 years old	90%		Private owners	4%
	25 – 30 years old	3%		Retired	0%
	31 -35 years old	1%		Unemployed	0%
	36 – 40 years old	0%	4.Educational Degree	Literate (Home-schooled)	0%
	41- 45 years old	4%		Primary school	0%
	46- 50 years old	2%		Junior high	0%
	51 – 55 years old	0%		Senior high	2%
	56 -60 years old	0%		Senior college/bachelor degree	90%
	61 -65 years old	0%		Master degree above	8%
	65 years old above	0%			

Among the 160 adjectives collected by the research team, 101 ones are left after removing the one with similar meanings. After making the word cards, 6 experts and scholars reduce them to 31 adjectives for the first time, after the mutual discussion and decision, the experts and scholars obtain 8 LOHAS' expected image words as shown in Table 5.

Table 5. The LOHAS' expected image words selected by experts

Life style ethnic group	Expected image vocabulary
LOHAS	Pragmatic, simple, humanistic, amiable, concise, organic, safe, convenient

Table 7. Descriptive analysis on the testees' basic information in the third stage

		Frequency	Percent
Gender	Male	16	69.6%
	Female	7	30.4%
Age	19–24 years old	17	73.9%
	31–35 years old	2	8.7%
	41–45 years old	2	8.7%
	46–50 years old	2	8.7%
Design specialty	Design background	22	95.7%
	Non-design background	1	4.3%
Educational background	Senior college/bachelor	13	56.5%
	Senior high and vocational school	1	4.3%
	Master degree above	9	39.2%
	Total	23	100.0%

Firstly, the research team uses key words to collect 400 pictures about the LOHAS from free authorized map depots of search engines, such as google.com (LIFE), picjubmo.com (Picjumbo), freeimages.com (Stock Exchange), imcreator.com (IM Free), etc. and make the 4 × 6-inch 300pdi picture card, then, the experts and scholars classify and reduce them into 20 pictures (as shown in Table 6) after sufficient discussion.

The third stage is about investigations on LOHAS and image vocabulary, LOHAS and picture cognition, and the two questionnaires are made among the same group of testees who are screened out as the representatives in the first stage and are willing to participate in the third stage, totally 23 persons (7 men and 16 women), female testees, and testees aged from 19 to 24 years old and with the design background are in the majority, and most of them have the senior college/bachelor degree (Table 7).

Table 6. Results of experts' picture screening (only several examples)

According to the questionnaire results, from the mode line-holding situation (as shown in Tables 8 and 9), the testees feel about LOHAS as being optimistic, bright, neither masculine or feminine, neither robust nor exquisite, active and quiet.

Table 8. The LOHAS' SD findings

Variable	Group	Frequency	Percent	Variable	Group	Frequency	Percent
Optimistic - Pessimistic	1	11	48%		1	2	9%
	2	5	22%	Exquisite - Robust	2	9	39%
Optimistic - Pessimistic	3	7	30%		3	11	48%
Bright - Shady	4	0	0%		4	1	4%
	5	0	0%		5	0	0%
	1	10	43%		1	5	22%
	2	7	30%		2	11	48%
Bright - Shady	3	5	22%	Active - Passive	3	4	17%
Masculine – Feminine	4	1	4%		4	3	13%
	5	0	0%		5	0	0%
	1	0	0%		1	0	0%
	2	3	13%		2	1	4%
	3	12	52%	Noisy - Quiet	3	5	22%
Masculine – Feminine	4	6	26%		4	9	39%
	5	2	9%		5	8	35%
					Total	23	100%

The matching results between the LOHAS' [pragmatic] expected image word and 20 pictures (as shown in Table 10), Figure 6 and Figure 17 are closest to the convenient image, and the closeness degree ranking is Figure 6 = Figure 17 > Figure 19 > Figure 3 = Figure 9 = Figure 16 > Figure 10 = Figure 13 > Figure 5 > Figure 1 = Figure 2 = Figure 8 = Figure 15 > Figure 4 = Figure 12 = Figure 14 = Figure 18 = Figure 20 > Figure 7 = Figure 11.

The matching results between the LOHAS' [simple] expected image word and 20 pictures, Figure 3 is closest to the convenient image, and the closeness degree ranking is Figure 3 > Figure 4 = Figure 6 = Figure 16 = Figure 17 = Figure 18 > Figure 8 = Figure 13 > Figure 5 > Figure 2 = Figure 9 = Figure 10 > Figure 1 = Figure 15 > Figure 7 = Figure 11 = Figure 12 = Figure 20 > Figure 14 > Figure 19.

The matching results between the LOHAS' [humanistic] expected image word and 20 pictures, Figure 4 is closest to the convenient image, and the closeness degree ranking is Figure 4 > Figure 6 > Figure 2 = Figure 8 = Figure 17 > Figure 9 = Figure 12 = Figure 13 = Figure 17 > Figure 3 > Figure 10 = Figure 16 > Figure 5 = Figure 20 > Figure 1 = Figure 18 > Figure 14 > Figure 7 = Figure 11 = Figure 15.

Table 9. Findings of the LOHAS's expected intentions

Category Item	Mode Interval Optimistic – Pessimistic	Bright Shady	- Masculine – Feminine	Exquisite - Robust	Active Passive	- Noisy - Quiet	Quantity of Complying Intervals	The Most Integrating Ranking
LOHAS	1	1	3	4	2	4	-	-
Pragmatic	3	2	2	2	2	2,3	1	8
Simple	1	2	3	3	4	5	2	4
Human-istic	1	1	3	2	1,2	3	4	2
Amiable	1	1	3	2	1	4	4	1
Concise	3	1	3	1	3	3	2	4
Organic	1	1	4	2	2	3	3	3
Safe	1	1	3	3	1	3	2	4
Conven-ient	2	2	2	2,3,4	2	3	2	7

Explanation: the mode in the findings is the majority's feelings, if more than 2 intervals have the same mode, it indicate the feelings are scattered.

Table 10. Matching results between expected image word and pictures of the LOHAS' [pragmatic]

Category Item	Mode Interval Optimistic – Pessi-mistic	Bright - Shady	Mascu-line – Feminine	Exquisite - Robust	Active - Pas-sive	Noisy - Quiet	Quantity of Complying Intervals	The Most Integrat-ing Rank-ing
Pragmatic	3	2	2	2	2	2,3	-	-
Figure 1	3	1,3	3,4	4	4	4	1	10
Figure 2	1	1	3	2	3	4	1	10
Figure 3	1,2	2	3	3	2	4	2	4
Figure 4	1	1	3	3	3	3	0.5	14
Figure 5	3	3	3	3	3	3,4,5	1.3	9
Figure 6	1	2	3	2	3	2,3	3	1
Figure 7	4	4	4	3	3,5	4	0	19
Figure 8	1	1,2	4	2	1,3	5	1	10
Figure 9	1,3	2	2,3	2	3	4	2	4
Figure 10	3	1,4	3	3	3	2	1.5	7
Figure 11	4	4	4	4	5	5	0	19
Figure 12	1	1	4	2,4	3	4	0.5	14
Figure 13	2	1,2	3	2	3	5	1.5	7
Figure 14	4	4	2,3	4	3	4	0.5	14
Figure 15	3	4	4	3,4	4	4	1	10
Figure 16	1	2	4	3	2	2	2	4
Figure 17	3	2	3	2	4	3	3	1
Figure 18	3,4	3	3	3	3	5	0.5	14
Figure 19	3	1	4	2	3	3,4	2.5	3
Figure 20	1,2,4	4,5	4	3	3	3	0.5	14

The matching results between the LOHAS' [amiable] expected image word and 20 pictures, Figure 2 is closest to the convenient image, and the closeness degree ranking is Figure 2 > Figure 12 > Figure 4 = Figure 6 = Figure 8 = Figure 3 = Figure 13 = Figure 19 > Figure 17 > Figure 1 = Figure 10 = Figure 14 > Figure 5 > Figure 7 = Figure 15 = Figure 16 = Figure 18 > Figure 20 > Figure 11.

The matching results between the LOHAS' [concise] expected image word and 20 pictures, Figure 4 is closest to the convenient image, and the closeness degree ranking is Figure 4 > Figure 10 = Figure 19 > Figure 5 > Figure 2 = Figure 17 > Figure 6 = Figure 13 = Figure 18 > Figure 1 = Figure 9 = Figure 12 = Figure 20 > Figure 14 => Figure 3 = Figure 8 = Figure 15 > Figure 7 > Figure 11 = Figure 16.

The matching results between the LOHAS' [organic] expected image word and 20 pictures, Figure 8, Figure 12 and Figure 19 are closest to the convenient image, and the closeness degree ranking is Figure 8 = Figure 12 = Figure 19 > Figure 2 = Figure 4 = Figure 16 > Figure 6 > Figure 17 = 20 > Figure 3 = Figure 9 = Figure 13 > Figure 1 = Figure 7 = Figure 11 = Figure 15 > Figure 10 > Figure 5 > Figure 14 = Figure 18.

The matching results between the LOHAS' [safe] expected image word and 20 pictures, Figure 4 is closest to the convenient image, and the closeness degree ranking is Figure 4 > Figure 2 > Figure 3 = Figure 6 = Figure 10 = Figure 16 = Figure 17 = Figure 18 > Figure 5 = Figure 20 > Figure 8 => Figure 12 = Figure 16 = Figure 17 = Figure 18 > Figure 13 = Figure 19 > Figure 7 = Figure 9 => Figure 1 = Figure 14 = Figure 15 > Figure 11.

The matching results between the LOHAS' [convenient] expected image word and 20 pictures, Figure 3 is closest to the convenient image, and the closeness degree ranking is Figure 3 > Figure 16 = Figure 17 > Figure 13 > Figure 20 > Figure 9 > Figure 4 > Figure 6 = Figure 8 = Figure 19 > Figure 12 = Figure 15 > Figure 5 > Figure 14 > Figure 1 = Figure 2 = Figure 7 = Figure 10 = Figure 11 = Figure 18.

5 Conclusions and Suggestions

According to the findings of the research, the following conclusions can be reached

1. The LOHAS are optimistic, bright, active and quiet.
2. The LOHAS' expected semantic vocabulary ranking is amiable > simple, concise, safe > convenient > pragmatic.
3. Regarding the matching situation between the LOHAS's expected vocabulary and pictures, Figure 6 and Figure 17 are closest to the pragmatic image; Figure 3 is closest to the simple image; Figure 4 is closest to the humanistic image; Figure 2 is closest to the amiable image; Figure 4 is closest to the concise image; Figure 8, Figure 12 and Figure 19 are closest to the organic image; Figure 4 is closest to the safe image; and Figure 3 is closest to the convenient image.
4. Figure 3 and Figure 4 are closest to the expected image.

As the research is initial, limited to time, only a simple life style ethnic group is researched, and it is suggested to investigate many ethnic groups which helps the interactive application during the imagine billboard design. Besides, with the advent of the big data era, a large-scale database is built for the life style ethnics, expected image vocabulary, and pictures to improve the design efficiency and consumer satisfaction.

Acknowledgments. I would like to thank the Ministry of Science and Technology of Taiwan for providing assistance in research funding and for all the people who participated in the study.

References

1. Jansson, D.G., Smith, S.M.: Design fixation. Des. Stud. **12**(1), 3–11 (1991)
2. Brown, T.: Change by Design: How Design Thinking Transforms Organizations and Inspires Innovation. Hardcover, New York (2009)
3. Pan, G.-Y., Zhang, Y.-M.: A Study of the Process of Mature Period Product Differentiation - The Proposal of Multi-image Features Deduction Method. National Cheng Kung University of Industrial Design, Taiwan, R.O.C. (2006)
4. Baxter, M.: Product Design. CRC Press, New York (1995)
5. Kobayashi, S.: Color Marketing Strategies. Japan Color Design Institute, Japan (1986)
6. Lan, Y.-R.: Design of "Yue Su" Toys. National Taiwan University of the Arts Institute of Art, Master of Arts, Unpublished. Taiwan, R.O.C. (2006)
7. Wikipedia. https://zh.wikipedia.org/wiki/. Accessed 03 May 2015
8. Dobni, D., Zinkhan, G.M.: In search of brand image: a foundation analysis. Adv. Consum. Res. **17**, 110–119 (1990)
9. Robbeets, M.: How the actional suffix chain connects Japanese to Altaic. Turkic Lang. **11**(1), 3–58 (2007)
10. Martin, R.A.: The Psychology of Humor: An Integrative Approach. Elsevier Academic Press, Burlington (2007)
11. Nagamachi, M.: Kansei engineering: a new ergonomic consumer-oriented technology for product development. Int. J. Ind. Ergon. **15**, 3–11 (1995)
12. Jindo, T., Hirasago, K., Nagamachi, M.: Development of a design support system for office chairs using 3-D graphics. Int. J. Ind. Ergon. **15**, 49–62 (1995)
13. Kelly, G.: Personal Construct Theory (1955)
14. Lazer, W.: Life style concept and marketing. In: Greyser, S. (ed.) Toward Scientific Marketing. American Marketing Association, Chicago (1963)
15. Osgood, C.E., Suci, G.J., Tannenbaum, P.H.: The Measurement of Meaning. University of Illinois Press, Urbana (1957)
16. Hofstattr, P.R.: Handbuch d. Neurosenlehre u. Psychotherapie, vol. I, Monaco-Berlino (1958)
17. Huang, J.-H.: Semantic analysis. In: Social and Behavioral Science Research, pp. 721–740. Tung Wah Bookstore Press, Taipei (1973)
18. Eastern Online Co., Ltd.: E-ICP, Eastern Integrated Consumer Profile, Taipei, Taiwan, R.O.C. http://www.isurvey.com.tw/3_product/1_eicp.aspx. Accessed 03 June 2015

Defining 'Architecture' for Software Engineering – A Review of Terminology

Julio C. Guzmán(✉), Gustavo López, and Alexia Pacheco

University of Costa Rica, San José, Costa Rica
{julio.guzman,gustavo.lopez_h,
alexia.pacheco}@ucr.ac.cr

Abstract. There are many terms used in the industry that relate to similar topics. Architecture is one of them. In many companies, representatives develop Information Architectures even though this term was conceptualized to be used in the domains of Information Systems or Information Technology. Moreover, different consultant agencies, consortia or working groups create standards or guidelines that sometimes use these concepts interchangeably. In this paper, we want to provide a comprehensible definition of the most common (Architecture-related) concepts used by members of the information technology community into non-software development companies. In addition, a literature review is also conducted to determine the academic use of these terms, and a website based review is conducted to gather the most common definitions provided in articles that do not undergo peer reviews.

Keywords: Enterprise architecture · Business architecture · Information architecture · Applications architecture · Technologies architecture · Solutions architecture · Infrastructure architecture · Systems architecture · Integration architecture · Software architecture

1 Introduction

Working in software engineering can be sometimes messy. The core concepts of software engineering are rooted in the senior software engineers. However, as younger people enter the domain, the definitions they hear vary drastically. The most common factors influencing the definitions of architecture include companies, industries, maturity of the workers and organizations, and even regionalization.

Nonetheless, the context in which architecture definitions are applied, forces people to bend the definitions themselves. In this paper, we collect the most common and used definitions for architectures in software engineering including: enterprise, business, information, solution, infrastructure, system and application, integration, technology, and software architectures.

The problems and risks of poor architectural practices are well known and documented within the software engineering profession. Even though, professionals try to avoid risks in software development by properly documenting architectures; the differences in terminology increase the risks in these projects and avoid proper communication.

T. Ahram and W. Karwowski (eds.), *Advances in Human Factors, Software, and Systems Engineering*, Advances in Intelligent Systems and Computing 598,
DOI 10.1007/978-3-319-60011-6_14

There are different organizations around the world, which offer forums, conferences and workshops to discuss the advances in software architecture and the core definitions; their publications offer different perspectives on each definition. However, the traditional sources define architectures in an extremely thorough and lengthy way, requiring a serious investment of time to fully digest them.

Our work tries to complement the definitions in an ordered and simple way. Our goal is to present a comprehensive definition of the most common architecture terms and represent their relationships. Our work's motivation relies on the importance of architectures in software engineering, as they are the primary factor in the creation and evolution of all software products.

This paper will be divided into four sections. Section 2 includes other authors' perspectives regarding the definition of all-ten named architectures. Section 3 will explain our perspective regarding our proposal of the definition of architecture. Finally, in Sect. 4, a discussion dwelling on the importance of having a common definition of architecture is described.

2 Perspectives

This section describes authors′ definitions for the different architectures. Henceforth, offering different approaches and perspectives for the concepts, and reviewing each one of them.

2.1 Enterprise Architecture

The Institute of Electrical and Electronics Engineers defines Enterprise Architecture as: "*An architecture is the fundamental organization of a system embodied in its components, their relationships to each other, and to the environment, and the principles guiding its design and evolution*" [1].

Yet, not everyone seems to adhere to this definition, and have also defined Enterprise Architecture as: "*...a coherent set of descriptions, covering a regulations-oriented, design-oriented and patterns-oriented perspective on an enterprise, which provides indicators and controls that enable the informed governance of the enterprise's evolution and success*" [2].

The Open Group's Architectural Framework (TOGAF) defines this architecture with two meanings depending on the context; for this research, the suitable definition of Enterprise Architecture is a group of "*... structure of components, their interrelationship, and the principles and guidelines governing their design and evolution*" [2].

The Gartner Group defines Enterprise Architecture as: "*the process of translating business vision and strategy into effective enterprise change by creating, communication, and improving the key principles and models that describe the enterprise's future state and enable its evolution*" [3].

2.2 Business Architecture

According to Simon and Schmidt, Business Architecture is "*a general description of a system. It identifies its purpose, vital functions, active elements, and critical processes and defines the nature of the interaction between them*" [4].

The Gartner Group defines Business Architecture as a part of Enterprise Architecture. For them, Business Architecture is the context where Enterprise Architecture is developed, and this context is a vision of the future state, an anchor model, and a set of guiding principles.

Using the Architecture Development Method ADM-TOGAF v9, "*Business Architecture is defined as a layer focused on strengthening business architecture and IT governance*" [5].

2.3 Information Architecture

The Information Architecture of any organization involves the definition of relationships between business processes and data classes. This architecture "*provides the basis for resource management and tactical planning, allowing for an orderly implementation of Information Architecture*" [6].

Other definition establishes Information Architecture as a discipline, in which critical thinking is promoted and grounded discussions in the community are improved. Information Architecture is inherently transdisciplinary, because data needs to be available to complete the different activities within the organization. So, information architecture is a mix of disciplines and models [7].

The Gartner Group defines Information Architecture as "*... all the sources of information—including paper, graphics, video, speech and thought—that define the enterprise...*" [8]; which is represented by this layer of applications architecture. "*... It also defines the sources and destinations of information, its flow through the enterprise, as well as the rules for persistence, security and ownership*" [8].

2.4 Solutions Architecture

The Gartner Group defined Solutions Architecture "*as an architectural description of a specific solution. SAs combine guidance from different enterprise architecture viewpoints (business, information and technical), as well as from the enterprise solution architecture*" [9].

The Solutions Architecture is defined simple like the organization of investments in solutions. These investments include training of architects, development of architectures and implementation of architecture processes.

2.5 Infrastructure Architecture

Infrastructure Architecture is in charge of "*... modularity, makes viable the independent evolution of each component, and allows the customization of the overall infrastructure...*" Moreover, the Infrastructure Architecture brings interoperability and scalability [10].

Gartner Group infrastructure defines the scope, scale and design necessary to provide applications services [11].

2.6 Systems Architecture

The Systems Architecture "*... are mainly composed of information systems*" [12]. This architecture should support changes in an efficient and effective manner.

Systems Architecture usually describes three separate models; "*the applications depiction, the hardware depictions showing the hardware configuration, and a communications depiction describing networks and hardware platforms where the data is movement throughout the enterprise*" [13].

2.7 Integration Architecture

Integration Architecture is defined as a vertical architecture because influents others architectures of Enterprise Architecture. The Integration Architecture "*define in the systems the parts and their connections in a systematic way with the aim to preserve maximum agility and minimize integration cost*" [14].

For The Gartner Group, the Integration Architecture is part of system integration and is "*view as a process of creating a complex information system that may include designing or building a customized architecture or application, integrating it with new ones or with existing hardware, packaged and custom software, and communications*" [15].

2.8 Applications Architecture

The Applications Architecture "*must provide the appropriate mechanism to represent every element in the conceptual schema, such as classes, attributes, relationships, derivations and so on*" [16]. In addition, the Applications Architecture specifies how each mechanism is implement and their interactions.

2.9 Technologies Architecture

The Technologies Architecture is the use of the information and communications technologies, these components are in service of the government by making it more accessible, effectible, and accountable. Also, it is based on functionality, complexity and maturity [17].

The Gartner Group defined the Technologies Architecture as a "*viewpoint where the reusable standards, guidelines, individual parts and configuration that are technology-related. Technologies Architecture defines are focus on the technical domains*" [18].

2.10 Software Architecture

Software Architecture is defining as "*a structure that organizes the software elements and the resources of a software system. Software elements and resources are represented by subsystems, these subsystems have specific responsibilities and relationships to other subsystems*" [12].

The Software Architecture details the behavior and the relationship of the subsystem, not the internal structure of the subsystems, and host enterprise applications.

3 Proposal

This section describes our proposal for architecture definitions including the Enterprise Architecture and its four perspectives: business, information, application and technology. Moreover, we delve into the definition of architecture models that actually represent built systems: system, infrastructure, solution and integration. Additionally, we include software architecture as another force that drives system development (Fig. 1).

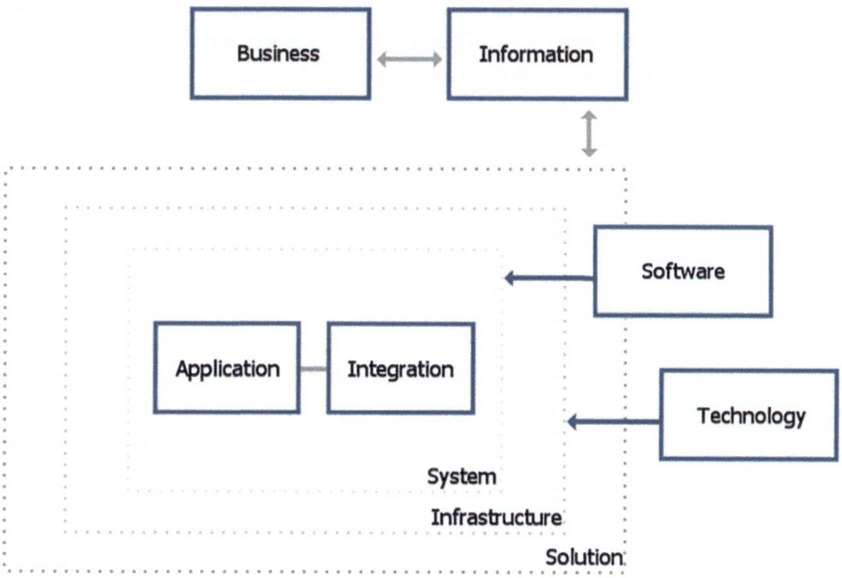

Fig. 1. Schematic representation of the enterprise architecture and their relations

3.1 Enterprise Architecture

The Enterprise Architecture is a conceptual representation that defines the structure and operation of an organization. The goal of an Enterprise Architecture is to provide the organization management and leaders with recommendations to achieve results, aligned with the goals of the organization.

The most adopted definition of Enterprise Architecture encompasses four perspectives: business, application, information and technology. These perspectives can also be architectures and they define how each part of the company will function and evolve. Traditionally, Enterprise Architectures are large documents that describe the organization's operating model. However, there are not many published Enterprise Architectures for various understandable reasons.

3.2 Business Architecture

Business Architecture is a model that encompasses an organization's goals, initiatives, projects, services, products, policies, and regulations. It also contains the tactical information: their vision, tactics and strategies.

Working in close relation with the Enterprise Architecture, the Business Architecture represents aspects of the business domains. Moreover, it represents how the business aspects and their interactions affect the organization.

The goal of the Business Architecture is to provide a competitive advantage to the organization, allowing in-time decision making. The basic components of this architecture are the business' objectives.

There are two parts of the Business Architecture. The first drives the business and helps decision makers to align their decisions with a centralized architecture. The second describes and defines the interaction between business strategies, processes and organizational information needs.

3.3 Information Architecture

Information Architecture is a more technical model when compared to Enterprise Architecture and Business Architecture. It usually represents the shared information in the organization, in a model.

The goal of the Information Architecture is to provide a framework that allows the labeling, categorization and classification of the data. This type of architecture is one of the most controversial ones, especially because it is used in several fields.

For our purposes, Information Architecture includes a guide that allows the flexible exchange of information between people. Besides, data is arranged in a more useful fashion for the organization´s collaborators.

3.4 Solutions Architecture

Solutions Architecture is a model that represents how the Infrastructure Architect works in order to resolve the business needs. The goal of the Solutions Architecture is to use available data and information to complete the organization´s activities. The different needs and capabilities of the business are proposed in Business Architecture and go through Information Architecture in order to be used as solutions.

3.5 Infrastructure Architecture

Infrastructure Architecture influences all applications and systems with the enterprise. Infrastructure Architecture depends on what Technology Architecture establishes.

The goal of Infrastructure Architecture is to provide a framework where systems will be developed. Infrastructure Architecture defines the growth of Systems Architecture because it sets up the base to support Systems Infrastructure.

3.6 Systems Architecture

Systems Architecture is a model that represents the reality of how software and systems were developed. The main flow of information is delivered by systems throughout all the enterprise.

The goal of Systems Architecture is to response to business needs, using in the best way possible the infrastructure. At the end, the Systems Architecture is a trusted representation of the applications used in the enterprise. If we desire to approach the architectures in a hierarchical model, Systems Architecture would be supporting both, Integration as well as Application Architectures, and therefore, being the face of the infrastructure.

3.7 Integration Architecture

Integration Architecture is a model that allows the articulation of the applications in the enterprise. It is an essential part between Systems Architecture and Applications Architecture, linking these two architectures.

The goal of the Integration Architecture is to manage the evolution of the Applications Architecture integrated into the Systems Architecture. The value produced by the different applications will be incremented while the applications are integrated while sharing critical information.

The proper management of the Integration Architecture guarantees the interoperability of the systems and the operations in charge of the business.

3.8 Application Architecture

Even though the traditional definition of application architecture is a model describing the behavior of applications and their interactions, this definition is more extensive than how application architectures are usually applied.

Application Architectures usually describe how a system is deployed, focusing especially on the processing capabilities of the system. It encompasses the application, interoperability capabilities, its characteristics for performance, scalability, and the application lifecycle. It is common to observe that application architectures focus on how data flows in the system rather than the actual blueprints of the system. There is a strong relationship between Application Architecture and Integration Architecture as the data flows through applications using the Integration Architecture.

When an organization builds systems that are independent, they usually omit Integration Architecture and they describe the integration features and characteristics in the application architecture.

3.9 Technology Architecture

In the Technology Architecture, organizations define standards and guidelines that unify the way in which the organization manages their technical domain. The technology application strongly affects infrastructure deployment. Likewise, Technology Architecture describes how infrastructure supports the organization's operations.

Some companies include in their technology application, technical characteristics, such as database management, operating systems, and application development standards. However, in our proposal this architecture describes the trends that affect infrastructure. The influence of Technology Architecture on infrastructure is homologous to the influence that Software Architecture has on the systems.

3.10 Software Architecture

Software Architecture is an organizational model of the elements and resources in the systems. It is necessary to make a difference between Software Architecture in the Enterprise Architecture context and Software Architecture in the software development context. Software Architecture is how elements serve the business.

Software Architecture is independent from Systems Architecture, but Software Architecture depends on the scope and support of Systems Architecture to represent how the applications use the systems.

4 Discussion

In the more pure concept, architectures are information. Each enterprise should defined there's own concept of architecture, context and framework where those architectures are developed, drawing the proper blueprint for each one of them. Business, information, applications and technology give a point of view of a tangible architecture, and the logical implementation give the levels of detail. For instance, Integration Architecture is the implementation of the information layer and Solutions Architecture is the point of view of the business needs.

It is necessary to set a common between enterprises and establish a homologous language in software engineering field. However, the scope of this homologous language can be high, as well as low.

This effort, try to connect software engineering language with enterprise language, coupling business processes integrated with technologies and ecosystems of systems substantiated in technologies. The process of define architectures is fractal and iterative, and each definition can change according with the requirements of the context, scope and systems. For example, a technological stack is able to connect processing, data storage, middleware and applications as basic components, but the interactions and interdependencies of the components change form system to system. In addition to that, it is normal feel overwhelmed by the concepts of the different architectures, so it is important stay focused on the scope on the level of detail.

The main value of the concepts presented in this research is the power of classification. Nowadays, it is not a well define line within software, information, systems, for mention some ones. With the suitable classification, it is possible approach the complexity of the architectures and their components. The traceability in Enterprise Architecture is the key to set standard language and ensure success in the business.

Acknowledgments. This work was supported by CITIC at University of Costa Rica, Grant No. 834-B4-412.

References

1. IEEE Recommended Practice for Architectural Description of Software-Intensive Systems IEEE-SA Standards Board (2000)
2. Op't Land, M., Proper, E., Waage, M., Cloo, J., Steghuis, C.: Enterprise Architecture, pp. 113–125. Springer, Heidelberg (2009)
3. Lapkin, A., Allega, P., Burke, B., Burton, B., Bittler, R.S., Handler, R.A., James, G.A., Robertson, B., Newman, D., Weiss, D., Buchanan, R., Gall, N.: Gartner Clarifies the Definition of the Term "Enterprise Architecture" (2008)
4. Simon, D., Schmidt, C. (eds.) Management for Professionals Business Architecture Management Architecting the Business for Consistency and Alignment. Springer, Heidelberg (2015)
5. Cabrera, A., Abad, M., Jaramillo, D., Gómez, J., Verdum, J.C.: Definition and implementation of the enterprise business layer through a business reference model, using the architecture development method ADM-TOGAF. In: Mejia, J., Munoz, M., Rocha, Á., Calvo-Manzano, J. (eds.) Trends and Applications in Software Engineering, pp. 111–121. Springer, Cham (2016)
6. Sá, F., Rocha, Á.: Information architectures definition – a case study in a portuguese local public administration organization. In: Rocha, Á., Correia, A., Wilson, T., Stroetmann, K. (eds.) Advances in Information Systems and Technologies, pp. 399–410. Springer, Heidelberg (2013)
7. Lacerda, F., Lima-Marques, M.: Information architecture as a discipline—a methodological approach. In: Resmini, A. (ed.) Reframing Information Architecture, pp. 1–10. Springer, Heidelberg (2014)
8. The Gartner Group: information architecture - Gartner IT Glossary. http://www.gartner.com/it-glossary/information-architecture/
9. The Gartner Group: Solution Architecture - Gartner IT Glossary. http://www.gartner.com/it-glossary/solution-architecture/
10. Ciuffoletti, A., Congiusta, A., Jankowski, G., Jankowski, M., Krajiček, O., Meyer, N.: Grid infrastructure architecture: a modular approach from CoreGRID. Presented at the 2008
11. Robertson, B.: Modeling Architecture and Infrastructure Planning: Domains to Patterns and Beyond. Enterprise Planning & Architecture Strategies (2002)
12. Weske, M.: Evolution of enterprise systems architectures. In: Weske, M. (ed.) Business Process Management, pp. 25–69. Springer, Heidelberg (2012)

13. Rood, M.A.: Enterprise architecture: definition, content, and utility. In: Enabling Technologies: Infrastructure for Collaborative Enterprises, pp. 106–111 (1994)
14. Murer, S., Bonati, B., Furrer, F.J.: Integration architecture. In: Murer, S., Bonati, B., Furrer, F.J. (eds.) Managed Evolution, pp. 79–128. Springer, Heidelberg (2010)
15. The Gartner Group: system integration - Gartner IT Glossary. http://www.gartner.com/it-glossary/system-integration/
16. Pastor, O., Molina, J.C.: Application Architecture. In: Pastor, O., Molina, J.C. (eds.) Model-Driven Architecture in Practice, pp. 261–263. Springer, Heidelberg (2007)
17. Liu, C., Huang, Y., Pan, Q.: A study on technology architecture and serving approaches of electronic government system. In: Chen, R. (ed.) Intelligent Computing and Information Science, pp. 112–117. Springer, Heidelberg (2011)
18. The Gartner Group: Enterprise Technology Architecture (ETA) - Gartner IT Glossary. http://www.gartner.com/it-glossary/enterprise-technology-architecture-eta/

Design of a Secure Location Based Service for Mobile Cloud Applications

K. Meenakshi Sundaram[1(✉)] and Navjeet Kumar[2]

[1] Faculty of Computing, Botho University, Gaborone, Botswana
meenaksji@gmail.com
[2] Department of Computer Science and Engineering,
Saveetha School of Engineering, Saveetha University, Chennai, India
tonunavjeet@gmail.com

Abstract. The impact of Location Based Services (LBS) in the mobile cloud paradigm has been huge. The location based services provide a great backbone to mobile cloud and ease the whole process of communication at a global level. Any location based cloud service is used in order to enhance the user's accessibility and make the service more user-friendly. Using a cloud base location system can improve the quality of the data storage. The objective is creating a location based mobile cloud application in android platform. This application will be able to geo-tag captured images and upload them onto a cloud. The images that are uploaded in the cloud will be sorted according to their locality. The images in the cloud which are in public mode will be visible to a user when he is in the same locality of the picture's geo tag.

Keywords: Location based services · Mobile cloud · Geo-tag

1 Introduction

Cloud computing is the service in which resources is given to the customer on an on-demand kind of premise using the internet. One of the key features provided by cloud computing is the reduction of the downtime and the waste on expenditure for servers and other computer equipment. A company or a user is only required to purchase the systems with minimum amount of hardware specifications requires to handle the maximum processing power for their system. In the conditions where the strain and traffic are highly variable, it will lead to wastage of money. For example, if the number of users using a certain cloud service increases suddenly, then the systems must be able to cope up with the increase in demand [1].

Apart from just the mobile cloud computing which provides support to the clients there are various non-mobile cloud services that act as the backbone or the servers. The LBS is an application for mobile device that can pinpoint the location and other details of the device. They are generally based on the queries provided by the users, which are usually simple questions about the location, they can also be automated.

Any location based service will require the following components: the software provided by the service provider, a working mobile network for transmitting and receiving, information and requests for the service, the content supplier has to offer the location based information to the end user, access to the end user's mobile device and a

T. Ahram and W. Karwowski (eds.), *Advances in Human Factors, Software, and Systems Engineering*, Advances in Intelligent Systems and Computing 598,
DOI 10.1007/978-3-319-60011-6_15

positioning device. All the permission to location based services are not automated unless specified by the user. That means the user will have to subscribe to the service if he wishes to use it. It can be done by introducing the LBS application and by accepting a request, terms and condition that will be enabled to know the device's location by allowing the service provider [2].

2 Review on Mobile Cloud Computing and Applications

A mobile cloud application is an application that is capable of accessing the Internet through various types of portable computing devices.

Both mobile cloud applications and mobile web applications are familiar in nature and also in functionality as they depend on servers for computation, storage capabilities and they can access the internet services. Even though, it's often stated that every cloud application is a web application to, it's also true that not every web application is a cloud application. Thus, we can say that not every mobile and web applications can be kept running on a virtual environment without particularly being modified or optimized for it. This is due to the fact that; web application has initially been composed to be run and to store information on committed servers arranged in server farms. Cloud applications, then again are composed just to live and work on virtual servers in a distributed, multi-tenant engineering and store the information on the cloud. Along these lines, mobile cloud and web applications are both altogether different from typical versatile applications. The mobile applications are developed such that they keep running on one and only specific sort of mobile device or platform, and they are just installed on those mobile devices. The challenge in building up a mobile application is that developer(s) must create multiple versions of a similar mobile application in an event they are to be used on various other platforms like iOS and Windows based devices. As the cloud application, cannot be downloaded, the developers just have to develop a single rendition for the mobile cloud application that can be run on any device with just a browser and a working Internet connection. The challenge here is to develop and manage the application programming interfaces that can make use of all the loosely coupled cloud services as effectively and efficiently as possible [3].

The MCC provides various business opportunities to the cloud service providers and also the network operators or internet service providers. The MCC's can be defined as "unified versatile assets are influenced by the rich handheld innovation of fluctuated mists and system advancements towards portability, stockpiling and unhindered usefulness which serves a large number of mobile phones in any place, the channel of Ethernet or the Internet is gone through the anytime with respect to the heterogeneous stages and situations in the compensation of the light which uses the pay-per-usage value" [4].

2.1 Advantages of Mobile Cloud

- Mobile devices enable portability; they allow the users to access the cloud service at anytime and anywhere.

- Each mobile device gets dedicated resources; its own storage capacity, processing, detecting, and power resources which are worthwhile [5].
- Mobile cloud services enhance the user experience; they allow access to the user's location and context related data [6].

3 Proposed System

The proposed system will be making use of the following information,

- CID or Cell ID
- Mobile Country Code (MCC)
- Mobile Network Code (MNC)
- Location Area Code (LAC)
- Base Station ID
- Radio Frequency Signal
- The relevant strength between the adjacent towers.

3.1 A Components of Location Based Mobile Cloud Model

The structure of client and the server is represented in the Fig. 1. It shows the flow of information from client to server.

Client. The client system performs tasks like

- Capturing images
- Geo-tagging images

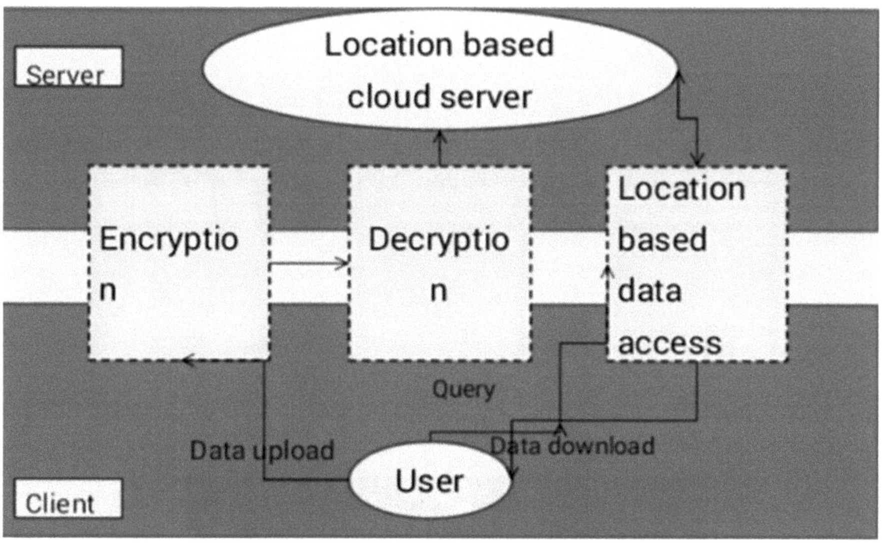

Fig. 1. Client and server model of secure location based mobile cloud

- Uploading images to cloud
- Downloading images from cloud
- Sending location data
- Sending queries

Not every client device will capture images and geo-tag them but every device will be providing the server with the details of its location. The location data will contain the MCC, MNC, LAC, CID, etc.

Server. The server will perform the following tasks,

- Computing location data
- Answering queries
- Downloading from client
- Uploading to cloud

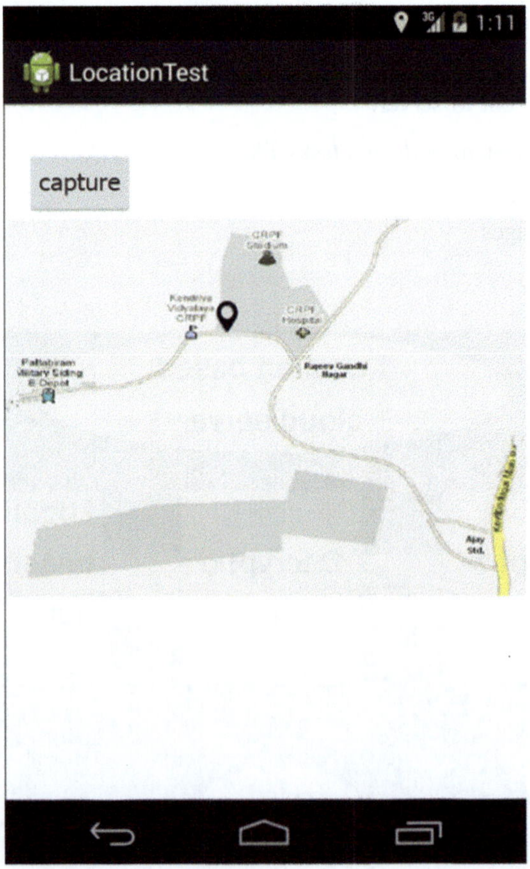

Fig. 2. Accessing the location data.

All the data will be encrypted while sending and receiving data in order to prevent anyone from getting hold of the location and the data produced by the client.

Location Based Server. In general, the LBS have the job of providing the services according to the client's location in real time which calls for background monitoring; it helps in triangulating the position. All this information will be interrelated with each other and be analyzed, which will be more efficient at pinpointing locations easily with less consumption of resources [11].

- GPS - The Global Positioning Satellites pinpoint the mobile phones location based on its signal and triangulates its location, and keeps track of it in real time.
- Cell Tower – this works almost like the Wi-Fi method, except here instead of a Wi-Fi station, there are cell transmitters and base [7].
- Wi-Fi – The mobile phone detects any nearby Wi-Fi signal and then compares the resulting pattern to an existing database of possible locations [8].

Fig. 3. Updating images to location directory.

All location based data is encrypted while sending and decrypted while receiving, this is necessary in order to keep the user's location secured and avoid leakage of private information [9].

In Fig. 2 as we can see that the device will pinpoint its location and nearby places, this is the initial state of the application and happens mostly on the client side along with the location server. If there are any existing photos in public mode available, they will be displayed.

In the second phase the device uploads the captured images onto the cloud server which will be collaborating with the location based server [12]. These images are essentially the pinpointing to the locations where they were captured at (Fig. 3). These locations can be common locations or even be heritage sites. Hence, the location can be explored by connected based on the topography and the landmarks [10].

4 Conclusion

In this paper, we have proposed a system for sharing pictures on a cloud platform which is location aware, it can help in identifying locations through pictures, and also promote tourism. This paper talks about the use of a location aware cloud server in the content sharing paradigm.

References

1. Shravanthi, C., Guruprasad, H.S.: Mobile cloud computing as future for mobile applications. Int. J. Res. Eng. Technol. **3**(5), 253–256 (2014)
2. Paelke, V., Dahinden, T., Eggert, D., Mondzech, J.: Location based context awareness through tag-cloud visualizations. In: The International Archives of the Photogrammetry, Remote Sensing and Spatial Information Sciences, vol. 38, Part II, pp. 290–295 (2012)
3. Shanklin, M.: Mobile cloud computing. http://www.cse.wustl.edu/~jain/cse574-10/ftp/cloud/#ToC
4. Zhu, Y., Ma, D., Huang, D., Hu, C.: Enabling secure location-based services in mobile cloud computing. In: 2nd ACM SIGCOMM Workshop on Mobile Cloud Computing, Hong Kong, China, pp. 27–32 (2013)
5. Hull, R.: Making scalable location-aware mobile apps, pp. 1–13. http://calvium.com/calvium/wpcontent/uploads/2012/09/Scalable_location_aware_apps_white_paper_Calvium.pdf
6. van Aart, C., Wielinga, B., van Hage, W.R.: Mobile cultural heritage guide: location-aware semantic search. In: Cimiano, P., Pinto, H.S. (eds.) Knowledge Engineering and Management by the Masses. EKAW 2010. Lecture Notes in Computer Science, vol, 6317. Springer, Heidelberg (2010)
7. Saroiu, S., Wolman, A., Enabling new mobile applications with location proofs. In: Proceedings of 10th Workshop on Mobile Computing Systems and Applications, HotMobile 2009, Santa Cruz, California, USA, 23–24 February 2009
8. Shetty, K.S., Singh, S.: Cloud based application development for accessing restaurant information on mobile device using LBS. Int. J. UbiComp (IJU) **2**(4), 37–49 (2011)

9. Debasish, J., Bandyopadhyay, D.: Controlled privacy in mobile cloud. In: 2nd IEEE International Conference on Recent Trends in Information Systems (ReTIS), 9–11 July 2015 (2015)
10. Calderoni, L., Maio, D., Palmieri, P.: Location-aware mobile services for a smart city: design, implementation and deployment. J. Theor. Appl. Electron. Commer. Res. **7**(3), 74–87 (2012)
11. Maa, X., Cuia, Y., Stojmenovicb, I.: Energy efficiency on location based applications in mobile cloud computing: a survey. J. Comput. **96**(7), 569–585 (2014)
12. Qi, Q., Cao, Y.: Cloud service-aware location update in mobile cloud computing. J. IET Commun. **8**(8), 1417–1424 (2014)

Ground Penetrating Radar for Measuring Thickness of an Unbound Layer of a Pavement

Rafiqul A. Tarefder(✉) and Mesbah U. Ahmed

Department of Civil Engineering, University of New Mexico, Albuquerque, NM 87131, USA
{tarefder,mahmed}@unm.edu

Abstract. Ground Penetrating Radar (GPR) is a widely used technology in the United States of America for measuring pavement layer thickness. GPR consists of an antenna that produces short duration electromagnetic pulses that penetrate pavement materials. The radar pulses are reflected at interfaces where the dielectric constants of material layers change. The reflected amplitude depends on the change in dielectric constant, while the arrival time of a reflected wave at a detector also depends on the depth at which the discontinuity is encountered. Layer material's dielectric property is used for pulse velocity and thickness calculation. Most of the GPR software performs the pulse velocity calculation using surface reflection method, where dielectric constant is assumed to be 4.0 for unbound materials, which may not be field representative. Simply, field moisture content may vary at different months over a year, which may lead to variations in the dielectric constant of an unbound layer. This study examines whether the default dielectric value, which is a common practice in USA, should be used for determining thickness of an unbound layer. In this study, using a percometer, the dielectric constant of an unbound layer material collected from an instrumented pavement section at the Interstate I-40, near Rio Puerco, New Mexico (milepost 141) was measured. GPR tests were conducted at the same instrumented section using 400 MHz and 900 MHz antennas. From analysis, it is observed that laboratory measured dielectric constant resulted in an accurate determination of layer thickness. Additional GPR and percometer tests were conducted on two other pavements: US 54 and US 285. Results confirmed that the accuracy level of the unbound layer thickness prediction drops significantly if the default dielectric constant value of 4.0 is used, as opposed to actual values. It is therefore recommended to use actual/measured dielectric constant values to determine unbound layer thickness using GPR technology.

Keywords: GPR · Dielectric constant · Unbound layer · Ground-coupled antenna · Moisture · Percometer

1 Introduction

A typical Ground Penetrating Radar (GPR) system has two major types of antennas. These are air-coupled and ground-coupled antennas [1–3]. Generally, an air-coupled antenna transmits high frequency signal such as 2.0 GHz whereas a Ground-Coupled Antenna (GCA) can transmits small frequency such as 900 MHz and below for

T. Ahram and W. Karwowski (eds.), *Advances in Human Factors, Software, and Systems Engineering*, Advances in Intelligent Systems and Computing 598,
DOI 10.1007/978-3-319-60011-6_16

pavement evaluation [4]. The high frequency antenna can only scan the asphalt layer at shallow depth due its high attenuation of signal energy. In this case, a GCA can scan unbound layers such as base and subgrade at deeper depth due to the small signal frequency [4].

During the analysis of GPR data, the signal velocity is required to determine layer thickness from the 2-way travel time [5]. A signal velocity depends on the Dielectric Constant (DC) of a medium through which it travels. The surface reflection method using a metal plate is conducted to determine the DC of a pavement top layer, i.e., asphalt layer. It is known as a velocity calibration method, which can only be performed using an air-coupled antenna. However, this velocity calibration method cannot be performed by a GCA. During the analysis of GCA data, the DC value is assumed to be equal 4.0, which may not be field representative [6]. In the field, moisture content varies in different months over a year, which may lead to high variation in the DC. It is necessary to measure the DC value of a specific unbound layer, and incorporate the measured value during the GCA data analysis.

To date, this study is initiated to investigate the effect of measured DC value on the accuracy of unbound layer thickness prediction from the GPR data whenever it is collected by a GCA.

2 Objectives

Main goal of this study is to investigate the accuracy of unbound layer thickness prediction from GPR data due to incorporation of measured dielectric constant. Specific objectives are to:

- Measure the dielectric constants of aggregates collected from base layers of selected pavement sections.
- Interpret base layer thickness from GPR data integrate these measured values as well as default value of 4.0.
- Compare the predicted thicknesses with as-built thickness to determine the accuracy level.

3 Test Background

3.1 GPR

A basic GPR system has two major components: transmitter and receiver (see Fig. 1(a)). During a single scan, transmitter shoots a signal towards the ground or pavement surface which travels through air and pavement layers. The transmitted signal is reflected back at different depths and these reflected signals are later captured by receiver. In a receiver, amplitude and 2-way travel time of the reflected signals at different depths are recorded to generate amplitude vs. time plot (see Fig. 1(b)). Here, the amplitude is in 'millivolt' (mV) and time is in 'nanosecond' (ns). This plot is typically known as 'A-scan'. In an 'A-scan', there are high amount of amplitude

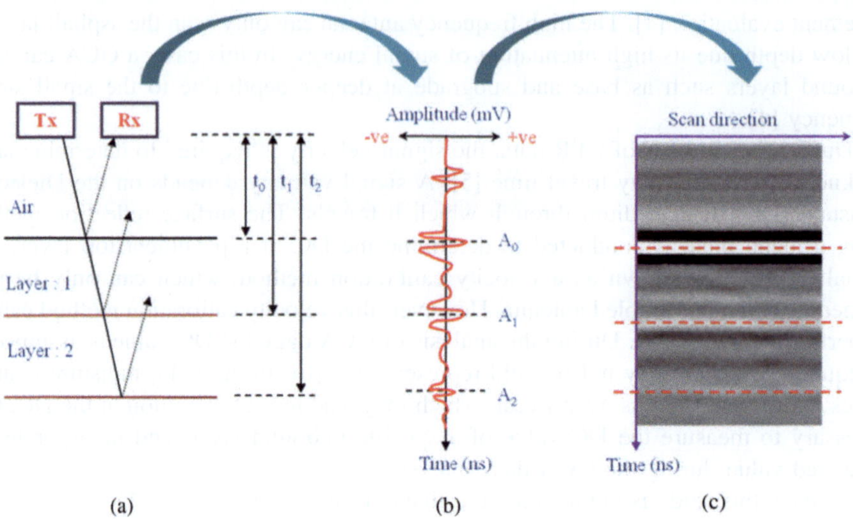

Fig. 1. GPR image during subsurface evaluation

variations at some depths. This type of amplitude variation is observed at some depth or interface wherever there is a material change.

During a GPR survey, a number of A-scans are recorded along the travel direction which is stacked to generate an image of subsurface (see Fig. 1(c)). This image is known as 'B-scan'. In a typical 'B-scan', the depth of layer interface, i.e., change in material, is identified based on color variation and contrast. For instance, the red dotted line at the center of the top 'black-white-black' color band with brightest contrast indicates an interface where the signal just enters the pavement surface from air. There are two more lines at deeper depths showing the presence of layer interfaces. In a conventional GPR data analysis, this simple technique is used identify pavement layers.

The two-way travel time at layer interfaces can be determined from Fig. 1(c). Once the time is determined, depth of a layer can be determined using the following relationship [5]:

$$h = \frac{vt}{2} \tag{1}$$

where $h =$ depth of layer, $v =$ velocity of signal in a specific media/pavement layer through which it travels, and $t =$ two-travel time of a signal. Signal velocity is dependent on dielectric constant of the media and can be determined as follows:

$$v = \frac{c}{\sqrt{\varepsilon}} \tag{2}$$

where $c =$ velocity of signal in air, and $\varepsilon =$ dielectric constant of a media. The dielectric constant can be determined from the layer velocity using the following relationship:

$$\varepsilon_l = \left(\frac{v_a}{v_l}\right)^2 \tag{3}$$

where ε_l = dielectric constant of layer. Once the dielectric constant of top layer is determined, the constants for the following layers can be determined as follows:

$$\varepsilon_n = \varepsilon_{n-1}\left[\frac{1-\left(\frac{A_0}{A_m}\right)^2+\sum_{i=1}^{n-2}\gamma_i\frac{A_i}{A_m}+\frac{A_{n-1}}{A_m}}{1-\left(\frac{A_0}{A_m}\right)^2+\sum_{i=1}^{n-2}\gamma_i\frac{A_i}{A_m}-\frac{A_{n-1}}{A_m}}\right] \tag{4}$$

where ε_n = dielectric constant of n-th layer, ε_{n-1} = dielectric constant of (n-1)-th layer, A_i = amplitude of signal reflected from i–th layer interface, and γ_i = reflection coefficient. The reflection coefficient can be determined as follows:

$$\gamma_i = \frac{\sqrt{\varepsilon_i}-\sqrt{\varepsilon_{i-1}}}{\sqrt{\varepsilon_i}+\sqrt{\varepsilon_{i+1}}} \tag{5}$$

where ε_i = dielectric constant of i-th layer, ε_{i-1} = dielectric constant of (i-1)-th layer, and ε_{i+1} = dielectric constant of (i + 1)-th layer. The dielectric constants, from the above mentioned relationships, are integrated to the GPR image for layer thicknesses prediction. This method is only applicable to air-coupled antenna. In case of GCA, this method cannot be used, and thereby, dielectric constant of 4.0 is assumed as a default value.

3.2 Percometer

Percometer is a nondestructive simple instrument to directly measure the Dielectric Constant (DC), ε, of any material at frequency of 40 Hz. During this test, a surface probe is just placed on a material surface and the device measures the DC by passing small amount of electricity over the thin layer of the test specimen (Fig. 2).

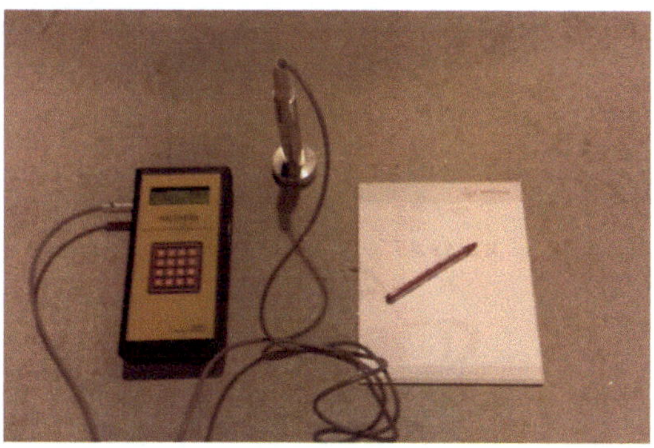

Fig. 2. Percometer test using a surface probe

4 Field and Laboratory Tests

4.1 Field

The GPR tests were conducted on three pavement sites, namely, I-40, US54, and US285 using a Geophysical Survey System Inc. (GSSI) system. The base layer thickness of these sites is 6 inch. During the test, four GCAs were operated simultaneously (see Fig. 3). These antennas were mounted at the rear of a truck. Among these, the middle two were 900 MHz whereas the rest were 400 MHz. Data were collected at a scan rate of 2/foot, and 512 traces/samples per scan.

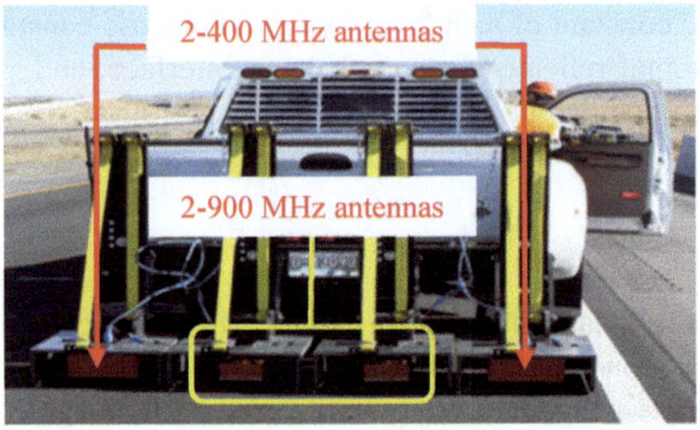

Fig. 3. GPR test with GCAs on I-40.

Aggregates from the base layer were also collected from these pavement sections for dielectric constant measurement in laboratory.

4.2 Laboratory

The methodology for sample preparation was followed according to the ASTM D1557-12 [7]. Test specimens with 4 in diameter and 6 in height were prepared by compacting 1 in successive layers under 25 blows by a proctor compactor. After sample preparation, dielectric constants of these specimens were measured using the percometer (Fig. 4).

Fig. 4. Dielectric constant measurement of field-collected aggregates (unbound material).

Some aggregates from the mixture were put in containers for moisture content measurement. Weights of the empty containers and with moist aggregates were measured. The containers were then put in an oven overnight to drive the moisture out from the aggregates. Dried aggregates were then weighed. Finally, MC (%) was determined as follows:

$$MC(\%) = \frac{w_{moist} - w_{dry}}{w_{dry} - w_{can}} \times 100 \quad (6)$$

where $MC(\%)$ = moisture content, w_{moist} = weight of container with moist soil, w_{dry} = weight of container with dry soil, and w_{can} = weight of empty container.

5 Results and Discussion

5.1 Dielectric Constant

The DC values measured at varying MC (%) of aggregates from the base layers are summarized in Table 1. DC increases with increase in moisture content since the water has a DC of 81. Base layer of the I-40 has a MC (%) of 8.0 which is the maximum among the three pavement sites. Due to this high MC (%), the DC is 12.0. The measured DC values are later incorporated to the GPR during the base layer thickness prediction.

Table 1. Dielectric constants of base layer

Site	MC (%)	DC, ε
I-40	8.0	12.0
NM47	5.5	7.6
NM14	6.5	9.1

5.2 Base Layer Thickness

The GPR data from different pavement sites are processed using the RADAN 7, a GPR post-processing software developed by the GSSI. The 2-way travel times at the top and bottom of the base layer of the I-40 were collected using the 400 and 900 MHz GCAs (Fig. 5(a)). The blue line represents the time corresponds to the top whereas the red line represents the bottom of the layer. The average time difference between the top and bottom of the base is 3.4 ns. During the thickness prediction from a GCA data, the RADAN 7 assigns a default DC of 4.0. Later, unbound layer thickness is determined as follows:

$$h = \frac{ct}{2\sqrt{\varepsilon}} = \frac{11.8t}{2\sqrt{\varepsilon}} \quad (7)$$

Here, DC (ε) is typically assigned as 4.0 by the RADAN 7. In this study, base layer thickness is also determined using the measured DC value. Figure 5(b) shows the

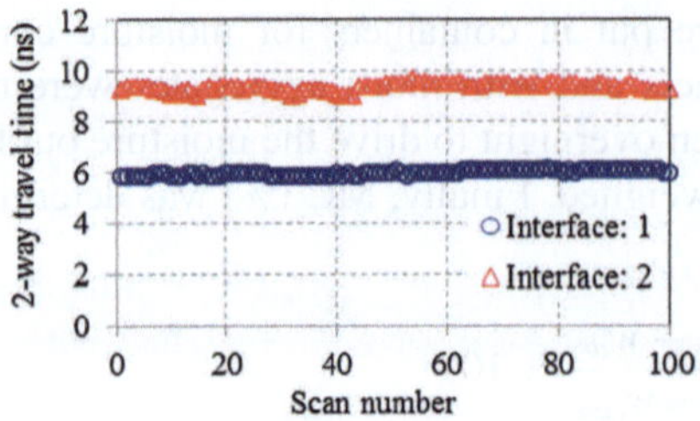

(a) 2-way travel time

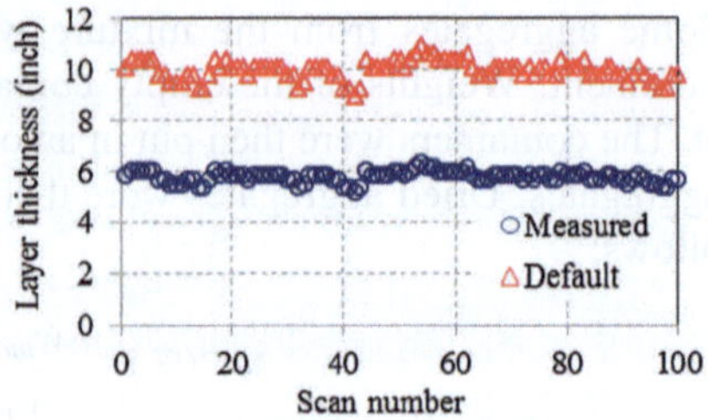

(b) Measured vs. Default

Fig. 5. Comparison of base layer thicknesses due to measured and default dielectric constant.

distribution of thicknesses due to measured and default DC values. It is evident that thickness due to measured DC is close to as-built thickness which is 6.0 inch.

The Root Mean Square of Error (RMSEs, %) based on these thicknesses are calculated as below:

$$RMSE(\%) = \sqrt{\frac{1}{n}\sum_{i=1}^{n}\left(\frac{T_{GPR} - T_{as-built}}{T_{GPR}}\right)^2} \times 100\% \tag{8}$$

where T_{GPR} = GPR predicted thickness, $T_{as-built}$ = as-built thickness, and n = number of GPR scans. The RMSEs (%) of predicted thickness due to measured and default DC are 5.5 and 66.4 respectively. It indicates that the incorporation measured DC values reduces the error significantly.

Figure 6 shows the distribution of the base layer thickness for both US54 N and US285 incorporating the measured DC. The thickness in US54 N is smaller than that in US285. In addition, thicknesses of both of these pavement sections are close to design thickness 6.0 inch. The RMSE (%) is calculated and these are 7.7 and 6.4 for US54 N and US285 respectively.

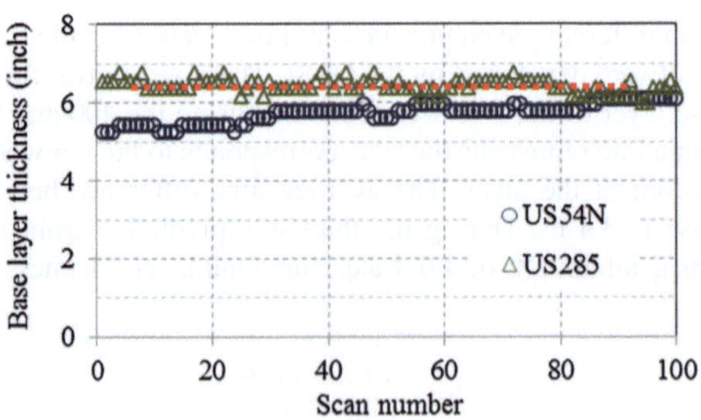

Fig. 6. Predicted base layer thicknesses (US54 and US285).

In summary, incorporation of measured dielectric constant improves the accuracy of the unbound layer thickness prediction from the GPR data collected by the ground-coupled antennas.

6 Conclusion

The study is performed to investigate the accuracy of the predicted layer thickness due to incorporation of a measured dielectric constant instead of default value whenever GPR data is collected using a ground-coupled antenna. Based on the overall observation, accuracy of base layer thickness prediction increases significantly. Finally, it is recommended to use measured dielectric constant values to determine unbound layer thickness using GPR.

Acknowledgments. This study is funded by the New Mexico Department of Transportation (NMDOT). The authors would like to express their sincere gratitude and appreciations to the Project Technical Panel Members, Project Advocate (Jeff Mann) and the Project Manager (Virgil Valdez) of NMDOT.

References

1. Saarenketo, T., Scullion, T.: Road evaluation with ground penetrating radar. J. Geophys. **43**, 119–138 (2000)
2. Maser, K.R.: Use of Ground-Penetrating Radar data for rehabilitation of composite pavements on high-volume roads. Transp. Res. Rec.: J. Transp. Res. Board **1808**, 122–128 (2002)
3. Al-Qadi, I.L., Lahouar, S.: Ground penetrating radar: state of the practice for pavement assessment. Mater. Eval. Am. Soc. Nondestr. Test. **62**(7), 759–763 (2004)
4. Ahmed, M.U., Tarefder, R.A., Maji, A.K.: Variation of FWD modulus due to incorporation of GPR predicted layer thicknesses. In: Proceedings of the 15th International Conference on Ground Penetrating Radar, Brussels, Belgium, June 30–July 4 (2014)
5. Willet, D.A., Rister, B.: Ground penetrating radar: pavement layer thickness evaluation. Research report: KTC-02-29/FR101-00-1F, Kentucky Department of Transportation (2002)
6. Geopghysical Survey Sytem Inc. (GSSI), RADAN7: Manual for software (2014)
7. ASTM D1557-12. Standard Test Methods for Laboratory Compaction Characteristics of Soil Using Modified Effort. American Standard for Testing and Material, 100 Barr Harbor Drive, PO Box C700, West Conshohocken, PA 19428-2959 (2012)

Research on Paper Submission Management System by Using Automatic Text Categorization

Qi Luo[1,2,3(✉)]

[1] College of Sports Engineering and Information Technology, Wuhan Sports University, Wuhan 430079, China
402135168@qq.com
[2] School of Educational Information Technology, Central China Normal University, Wuhan 430079, China
[3] Flat 5, La Medidionale, Gorg Borg Olivier Street, St. Julian's 1017, Malta

Abstract. The submission management system is a software system, also known as submission processing that streamlines and eases out the collection, tracking and management of paper submissions. Paper information can be received, authenticated, tracked, stored, and distributed electronically. The submission system of transaction and conference need sort thousands of papers into hundreds of reviewers to examine and comment. It is not good that these papers are sort into hundreds of reviewers to examine and comment in a short time, because the matching precision is low. Aiming at solving the above questions, the paper submission management system by using automatic text categorization is proposed in the paper. The system can automatically sort the submission papers and the published papers of the reviewer into the right subject, so we can judge the subject of the submission papers and the research direction of the reviewer.

Keywords: Automatic text categorization · Max frequency · Paper submission management system · Vector space model · Support vector machines

1 Introduction

Automatic text categorization refers to the task of sorting a set of documents into categories automatically from a predefined set. As the key technology in organizing and processing large amount of document data, text classification can solve the problem of information disorder to a great extent, and is convenient for users to find the required information quickly. Moreover, text classification has the broad applied in future as the technical basis of information filtering, information retrieval, and search engine.

The submission system of transaction and conference need sort thousands of papers into hundreds of reviewers to examine and comment. It is not good that these papers are sort into hundreds of reviewers to examine and comment in a short time, because the matching precision is low. Especially, when we do not know the research subject that the reviewer is well up, we cannot collect the information about the research

T. Ahram and W. Karwowski (eds.), *Advances in Human Factors, Software, and Systems Engineering*, Advances in Intelligent Systems and Computing 598,
DOI 10.1007/978-3-319-60011-6_17

subject that the reviewer is well up in time and accurately. These will baffle sorting the papers into categories normally. How to choose the proper reviewer to examine and comment paper is the key to value papers' quality rightly and step up the status of transaction and learning conference. So how to carry it out that the computer can automatically sort the papers to reviewer who is well up in the papers' subject. Automatic text categorization can solve the problem properly.

Aiming at solving the above questions, the paper submission management system by using automatic text categorization is proposed in the paper. The system can automatically sort the submission papers and the published papers of the reviewer into the right subject, so we can judge the subject of the submission papers and the research direction of the reviewer. Then the system can sort the submission papers to the reviewer who has the matching subject with the submission papers, we can also build a automatic recommended model for reviewer by the system.

2 Paper Submission Management System

The submission management system is a software system, also known as paper submission processing, that streamlines and eases out the collection, tracking and management of paper submissions. Paper information can be received, authenticated, tracked, stored, and distributed electronically. Submission management systems can be web-based system operating in a browser environment, a COTS based product, or may also be in the form of a desktop application. Paper submissions are completed electronically creating an efficient real-time process that saves time for both the submitter and recipient. Usually a paper submission management system can take in a high volume of data at fast rate.

A paper submission management system may be regarded as an application-specific content management system. In essence, such a system can be used an alternative to email in many situations.

Although paper submission management systems can get pretty complex, there are some features that a paper submission management system must provide in order for it to be classified under that category. Here is a list of some of those features:

- Provision to provide a start time and an end time for collecting paper submissions
- Mechanism for collection and storage of paper submissions
- Mechanism for downloading and viewing the collected paper submissions only to authorized people at any given time
- Mechanism to verify all documents and attachments are virus free
- Mechanism to guarantee all the required information is present for submission
- Mechanism for accepting and rejecting paper submissions
- Provision for giving relevant feedback to paper submitters
- Provision for requesting resubmission of paper submissions
- Mechanism for submitters to view paper submissions made by them and associated comments.

3 Document Categorization

The Document classification or document categorization is a problem in library science, information science and computer science. The task is to assign a document to one or more classes or categories. This may be done manually or algorithmically. The intellectual classification of documents has mostly been the province of library science, while the algorithmic classification of documents is mainly in information science and computer science. The problems are overlapping, however, and there is therefore interdisciplinary research on document classification.

The documents to be classified may be texts, images, music, etc. Each kind of document possesses its special classification problems. When not otherwise specified, text classification is implied.

Documents may be classified according to their subjects or according to other attributes (such as document type, author, printing year etc.). In the rest of this article only subject classification is considered. There are two main philosophies of subject classification of documents: the content-based approach and the request-based approach.

3.1 Content-Based Versus Request-Based Classification

Content-based classification is classification in which the weight given to particular subjects in a document determines the class to which the document is assigned. It is, for example, a common rule for classification in libraries that at least 20% of the content of a book should be about the class to which the book is assigned. In automatic classification it could be the number of times given words appears in a document.

Request-oriented classification is classification in which the anticipated request from users is influencing how documents are being classified. The classifier asks him: Under which descriptors should this entity be found and think of all the possible queries and decide for which ones the entity at hand is relevant.

Request-oriented classification may be classification that is targeted towards a particular audience or user group. For example, a library or a database for feminist studies may classify index documents differently when compared to a historical library. It is probably better, however, to understand request-oriented classification as policy-based classification: The classification is done according to some ideals and reflects the purpose of the library or database doing the classification. In this way it is not necessarily a kind of classification or indexing based on user studies. Only if empirical data about use or users are applied should request-oriented classification be regarded as a user-based approach.

3.2 Classification Versus Indexing

Sometimes a distinction is made between assigning documents to classes (classification) versus assigning subjects to documents. But as Frederick Wilfrid Lancaster has argued, this distinction is not fruitful. These terminological distinctions, he writes, are quite meaningless and only serve to cause confusion. The view that this distinction is purely superficial is also supported by the fact that a classification system may be transformed into a thesaurus and vice versa. Therefore, is the act of labeling a document (say by assigning a term from a controlled vocabulary to a document) at the same time to assign that document to the class of documents indexed by that term (all documents indexed or classified as X belongs to the same class of documents) [1, 2].

3.3 Automatic Document Classification (ADC)

Automatic document classification tasks can be divided into three sorts: supervised document classification where some external mechanism (such as human feedback) provides information on the correct classification for documents, unsupervised document classification (also known as document clustering), where the classification must be done entirely without reference to external information, and semi-supervised document classification, where parts of the documents are labeled by the external mechanism. There are several software products under various license models available [3].

3.4 Applications

Classification techniques have been applied to:

- Spam filtering, a process which tries to discern E-mail spam messages from legitimate emails
- Email routing, sending an email sent to a general address to a specific address or mailbox depending on topic
- Language identification, automatically determining the language of a text
- Genre classification, automatically determining the genre of a text
- Readability assessment, automatically determining the degree of readability of a text, either to find suitable materials for different age groups or reader types or as part of a larger text simplification system
- Sentiment analysis, determining the attitude of a speaker or a writer with respect to some topic or the overall contextual polarity of a document
- Article triage, selecting articles that are relevant for manual literature curation, for example as is being done as the first step to generate manually curated annotation databases in biology.

4 Paper Submission Management System by Using Automatic Text Categorization

First, in feature filtration, the paper firstly introduces the concept of Max Frequency and the correlation coefficient $D(m_{ik})$ between the item and the sort, and then puts forward the algorithm of improved χ^2. The experiment results show that the algorithm of Improved χ^2 performs well in feature filtration.

Second, for the item in an automatic recommended model for reviewer is the key word by using automatic text categorization, this paper puts forward the algorithm of vector space model and improves the algorithm of SVM. The experiment results show that, this algorithm can improve the quality of categorization model.

Third, the SVM categorization algorithm has a problem that the categorization speed will be slower with the increase of number for the categorization in many sorts. To solve the problem, this paper improves the SVM categorization algorithm. Results show that the improved SVM categorization algorithm can increase the process of alternation and make the improved SVM categorization learn by itself. In addition, the improved SVM categorization can sort the rate of sort accurately and accelerate.

5 SVM

In machine learning, support vector machines (SVM, also support vector networks are supervised learning models with associated learning algorithms that analyze data used for classification and regression analysis. Given a set of training examples, each marked as belonging to one or the other of two categories, an SVM training algorithm builds a model that assigns new examples to one category or the other, making it a non-probabilistic binary linear classifier. An SVM model is a representation of the examples as points in space, mapped so that the examples of the separate categories are divided by a clear gap that is as wide as possible. New examples are then mapped into that same space and predicted to belong to a category based on which side of the gap they fall [4].

In addition to performing linear classification, SVM can efficiently perform a non-linear classification using what is called the kernel trick, implicitly mapping their inputs into high-dimensional feature spaces.

When data are not labeled, supervised learning is not possible, and an unsupervised learning approach is required, which attempts to find natural clustering of the data to groups, and then map new data to these formed groups. The clustering algorithm which provides an improvement to the support vector machines is called support vector clustering and is often used in industrial applications either when data are not labeled or when only some data are labeled as a preprocessing for a classification pass.

More formally, a support vector machine constructs a hyper plane or set of hyper planes in a high- or infinite-dimensional space, which can be used for classification, regression, or other tasks. Intuitively, a good separation is achieved by the hyper plane that has the largest distance to the nearest training-data point of any class, since in general the larger the margin the lower the generalization error of the classifier.

Whereas the original problem may be stated in a finite dimensional space, it often happens that the sets to discriminate are not linearly separable in that space. For this reason, it was proposed that the original finite-dimensional space be mapped into a much higher-dimensional space, presumably making the separation easier in that space. To keep the computational load reasonable, the mappings used by SVM schemes are designed to ensure that dot products may be computed easily in terms of the variables in the original space, by defining them in terms of a kernel function $K(\mathrm{x}, \mathrm{y})$ selected to suit the problem. The hyper planes in the higher-dimensional space are defined as the set of points whose dot product with a vector in that space is constant. The vectors defining the hyper planes can be chosen to be linear combinations with parameters a_i of images of feature vectors x_i that occur in the data base. With this choice of a hyper

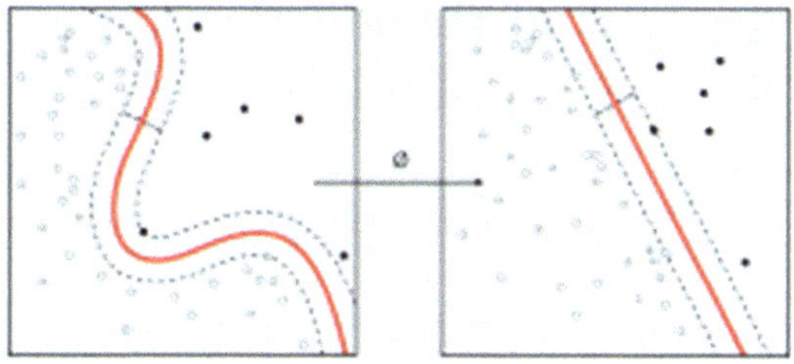

Fig. 1. Kernel machine

plane, the points x in the feature space that are mapped into the hyper plane are defined by the relation: $\sum_i a_i k(x_i, x) = cons\tan t$. Note that if $k(x, y)$ becomes small as y grows further away from x each term in the sum measures the degree of closeness of the test point x to the corresponding data base point x_i. In this way, the sum of kernels above can be used to measure the relative nearness of each test point to the data points originating in one or the other of the sets to be discriminated. Note the fact that the set of points x mapped into any hyper plane can be quite convoluted as a result, allowing much more complex discrimination between sets which are not convex at all in the original space [5, 6]. The Kernel machine is Fig. 1.

6 Conclusion

The paper puts forward the paper submission management system by using automatic text categorization. Through the automatic text categorization, the system can automatically sort the submission papers and the published papers of the reviewer into the right subject, so we can judge the subject of the submission papers and the research direction of the reviewer. Then the system also can sort the submission papers to the reviewer who has the matching subject with the submission papers. We can build a automatic recommended model for reviewer by the system.

References

1. Sebastiani, F.: Machine learning in automated text categorization. ACM Comput. Surv. **34**(1), 1–47 (2002)
2. Rossi, R.G., Lopes, A.d.A., Rezende, S.O.: Optimization and label propagation in bipartite heterogeneous networks to improve transductive classification of texts. Inf. Process. Manag. **52**(2), 217–257 (2016)
3. Aitchison, J.: A classification as a source for thesaurus: the bibliographic classification of H. E. Bliss as a source of thesaurus terms and structure. J. Documentation **42**(3), 160–181 (1986)
4. Cortes, C., Vapnik, V.: Support-vector networks. Mach. Learn. **20**(3), 273–297 (1995)
5. Ben-Hur, A., Horn, D., Siegelmann, H., Vapnik, V.: Support vector clustering. J. Mach. Learn. Res. **2**, 125–137 (2001)
6. William, H., Teukolsky, S.A., Vetterling, W.T.: Numerical Recipes: The Art of Scientific Computing. Cambridge University Press, New York (2007)

Impact Analysis of Demonetization on Service Sector: An Indian Experience

Uma Narain[1(✉)] and R.K. Pattnaik[2]

[1] Jyoti Dalal School of Liberal Arts, NMIMS University, Mumbai 400056, India
uma.narain@nmims.edu
[2] S.P. Jain Institute of Management and Research, Mumbai 400058, India
rk.pattnaik@spjimr.org

Abstract. Late evening on 8th November 2016, the Government of India demonetized the high denomination bills of ₹ 500 and ₹ 1000 and sent shock waves across the nation. ₹ 14.5 trillion (i.e. 86% of currency), thus sucked out of the system, was to be deposited/expended within a specified time and eventually replaced by new ₹ 500 and ₹ 2000 denomination bills. The common man bore the brunt under the fallacy that the inconvenience inflicted was short term and in the long run, the extreme action of the government was aimed at a larger public good. This is a study of the purpose, the process and the impact of this mega exercise on service sector through a natural experiment.

Keywords: Demonetization · Indian economy · Reserve Bank of India · Currency-management · Service-sector · Cash less economy · Government as service provider

1 Introduction

Services sector, operating on cash transactions, comprises a large chunk of flourishing economy in India. The unorganized sector accounts for 48% of the gross value added (supply side of the economic growth) and 82% of the employment [RBI report: 2017]. Services account for a dominant share of informal work force (84.7%) across all enterprises outside of public and private corporate sector [1]. Characterized by low literacy, low banking penetration, uncertain employment opportunities and abject poverty, this overwhelmingly large sector is dependent on cash economy where unlike payment through banks, money rotates several times in a day. Late evening on November 8, 2016, Indian government sent shock waves across the nation by taking the extreme step of withdrawing the legal tender status of bills of ₹ 500 and ₹ 1000, comprising around 86% of all the cash in circulation [2]. Referred to as 'demonetization', these two bills were to be deposited in the banks within seven weeks while cash withdrawals were limited. Earlier experience of demonetization during 1946 and 1978 was miniscule and inconvenience negligible, compared to the current seismic tremors [3]. This time, the 'surgical strike' affected almost all stakeholders in the economy and halted practically every economic activity. The service sector bore the brunt, as it is entirely dependent on cash transactions. Withdrawal of 86% currency notes had a felt impact of almost 95% on the people operating in the informal sector.

T. Ahram and W. Karwowski (eds.), *Advances in Human Factors, Software, and Systems Engineering*, Advances in Intelligent Systems and Computing 598,
DOI 10.1007/978-3-319-60011-6_18

According to official sources, the aim of the demonetization was to curb corruption, counterfeiting, terrorist activities and to catch the accumulation of 'black money' i.e. income not declared to the tax authorities, mostly by high income group individuals, thereby preventing it from being spent on common public good. An indirect aim was to eventually make everyone part of the non-cash digital economy.

Following the demonetization move, there has been unprecedented criticism and debate at various fora, particularly after the government communication to all via a smartphone app, interestingly called 'Narendra Modi' (the Prime Minister). This message conveyed a survey report about 90% success of the demonetization program! During the initial weeks, there was no clarity on 'why' and 'how' of demonetization as both the rich and the poor were equally confused. By end of the first month, 77% cash in circulation had returned to the bank negating the assumption of black money. Anecdotal evidence surfaced about middle class and lower middle class being the worst hit as they conducted 98% transactions in cash; many did not have a bank account and lived on cash income. Many stood in the long queue outside banks and ATM machines for days, waiting to exchange the banned bills, some at the behest of their employers. The hardship resulted in over 100 cases of death across India. With no solution in sight, liquidity squeeze was hurting even sectors like real estate, gems and jewelry, cement, steel, consumer durables and the retail in a big way.

The political leadership continues to draw mileage out of the measure. The Central Bank (Reserve Bank of India) has been a mute spectator, only seen revising/updating the FAQs on its own website nearly 60 times! Today, four months after the pandemonium, there is no clarity on the amount of 'black money' seized as nearly 90% currency is supposed to have returned to the banks. So where had the black money gone? The lower growth in the following financial year is almost a certainty now. The official version continue to be optimistic about demonetization, show casing it as an exercise in long-term benefit and acknowledging that authorities were cognizant of the short-term costs. Subsequently, the government took several measures to lessen the hardship of the people: payment of fees in schools was differed; toll tax on highways and bridges was waived-off, withdrawal limit of cask from banks was slowly relaxed and use of old bills was allowed at hospitals and gas stations etc.

Against the above backdrop, the authors want to study this phenomenon in immediacy. This paper attempts to discuss the impact of demonetization in a wider context. It first looks at macro level to study the role of RBI as a custodian of money market and money management followed by assessment of the impact of demonetization on the fake currency, black money and Indian economy with special reference to service sector. At the micro level, the paper describes the natural experiment conducted on lower middle income and the poor people in Mumbai. We selected this group for two reasons: this group was not even the intended target of demonetization yet faced unprecedented hardship and therefore needed consideration; and because they comprised the service sector that purely operates on cash transactions. The last part deals with inference drawn from the natural experiment and the road ahead.

2 Money, Monetization and Demonetization

It is necessary to understand the three terms repeatedly used in current scenario and understand what they portend. There is no unique definition of **'money'** - it is a means of payment and thus a lubricant that facilitates exchange. Money also acts as a storage of value and a unit of account. In the real world, money provides monetary services along with tangible remuneration. It is for this reason that money has to have relationship with the activities that economic entities pursue. Therefore, for policy purposes, money is the set of liquid financial assets and any variation in the stock can influence aggregate economic activity.

Monetization is the process through which, money effectively serves as a medium of exchange (transaction demand for money) and storage of value (precautionary demand and speculative demand for money). This is facilitated through issuance of currency of various denominations by the Central Bank (Reserve Bank of India), referred to as cash transactions. Cash remains an important mode of payment in the Indian economy, especially in informal sector. Cash demand tends to increase in the beginning of the month when salaries are spent and tapers off at the end of the month when consumer spending returns it to business accounts. Similarly, currency seasonality mirrors the seasonality in economic activity. The Currency-GDP ratio has increased over years and is currently at more than 11% [World Bank, 2012].

Demonetization connotes withdrawal of currency in circulation making it legally ineffective for any transaction. In the present case, demonetization means that the legal tender character of the existing bills in denominations of ₹ 500 and ₹ 1000 issued by the Reserve bank of India till November 8, 2016 stand withdrawn. Consequently, no one can use these bills for business transactions and/or store of value for future usage.

2.1 Operational Framework/Practice

Demonetization is an extreme step, it neither happens in a vacuum nor without purpose. The RBI is entrusted with the task of managing the 'legal tender' of the country. 'To monetize or not to monetize' is best understood in the context of RBI Act 1934 section 24 (2).

Legal Provision for Currency Management. As per the preamble to the Reserve Bank of India Act, 1934 and the specific provisions of Section 3 and 22, RBI is the sole currency management authority for the issuance of bills. As per the provisions of Section 26, on the recommendation of the Government of India, by notification in the Gazette of India, RBI may declare bills of any denomination to cease to exist as legal tenders. Thus, the decision of GOI to withdraw the ₹ 500 and 1000 should have had the recommendation of RBI board. The silence of the RBI all through the trials and tribulations of public has raised serious questions on currency management.

Currency Management Architecture (CMA). RBI manages the currency through 19 issue offices and 4,075 currency chests and 3,746 small coin depots. The Reserve Bank places indent for the bills with printing presses based on an econometric model factoring in inter alia, real GDP growth prospects, rate of inflation and denomination wise disposal rate of soiled bills. At the time of the recent demonetization, there does not seem any preparedness on the part of RBI to store enough bills to handle the crisis. As against an indent of ₹ 23.9 billion for the year 2015–16, the total number of bills supplied was only ₹ 21.2 billion. Knowing this short fall, 86% bills (by value) were going to be demonetized, ignoring the laws of the market.

Challenges to the Operational Framework. RBI faces various operational challenges. One, India is the largest producer and consumer of hard currency, next only to China. The demand is high as hard currency continues to be the dominant means of payment with the circulation (in value terms) placed at ₹16.63 trillion as of end of March 2016. Second, government presses print 40% bills circulated in India and the remaining 60% emanate from Reserve Bank of India currency presses. However, the Reserve Bank of India largely depends on imports for the main ingredients required for printing these bills. When demonetization crippled availability of bills, the government presses, dependent on imports, could not print enough to meet the demand. Third, the change in the size of the new bills forced major structural changes in the machines, further crippling the supply line and aggravating the woes of the public. Fourth, the challenge of cleaning the system of counterfeit currency, estimated to be .0018% of the total currency in circulation, is a daunting task for RBI [4]. While demonetized currency lying in the banks is yet to be verified, the process could capture 3.4% of the face value of the counterfeit bills [5].

2.2 Current Debate

The Economic survey of the Government of India, released recently, observes that demonetization has the potential to generate long-term benefits in terms of reduced corruption, greater digitalization of the economy, increased flows of financial savings, and greater formalization of the economy. All these could eventually lead to higher GDP growth, better tax compliance and greater tax revenues. As the Economic survey further points out, demonetization in India is unprecedented in international economic history, in that it combines secrecy and suddenness amidst normal economic and political conditions. All other sudden demonetizations have occurred in the context of hyperinflation, wars, political upheavals or other extreme circumstances.

The government stance continues to be positive – as per the claim, the inconvenience is short-lived and the need for forbearance will last for 50 days only. Finance Minister has claimed that the measures taken would have a positive impact on the economy in the long and medium term: “A lot of money that operates in the shadow economy will now become a part of the banking structure itself. Banks will have a lot more money to support the economy. Private sector investment, which was so far lacking, will now get back into the economy. The banks, struggling because of the bad loan problem will have a lot more money to lend for agriculture, infrastructure sector, social sector, trade and industry….The possibility of banks having low-cost funds being available now to lend at lower cost also rises. Additionally, when more and more transactions will come into the banking network, in the long run, you will find the taxation, both direct and indirect, improving” [6].

Globally respected economist and former Prime Minister Man Mohan Singh has commented that the deaths of people and distress among the poor farmers and small traders has convinced him that the demonetization plan amounts to "organized loot" and "legalized plunder". He has called the process a "monumental mismanagement" that will shave at least 2 percentage points off India's economic growth. "Even those who say that this measure will do harm or cause distress in the short term but will be in the interest of the country in the long run should be reminded of what John Keynes said once, 'In the long run all of us are dead'." [7]. While commenting on demonetization, one of the former governors of RBI, Y.V Reddy has stated that the "institutional identity of the RBI has been damaged" and it is facing "reputational risk" [8]. Interestingly, eminent economists have not endorsed demonetization. Harvard professor, Nobel Laureate, Amartya Sen compares curbing the black money to Napolean's invasion of Russia [9].

3 Impact Analysis

3.1 Massive Decline in Currency in Circulation

The immediate impact of demonetization is on currency in circulation. During the post demonization period, the available data indicates that on December 16, 2016, currency in circulation was ₹ 9582.3 billion as against ₹ 16, 634.6 billion in March 31, 2016, recording a massive decline of 42.4%. On a y-o-y basis, such decline was 38.9%. This implies non-availability of cash for transaction purposes during the demonetization process. As reported in the media, non-availability of cash coupled with lack of technological readiness adversely affected the confidence of business both small and big, particularly agriculture and services.

3.2 Economic Growth

The impact on economic growth due to demonetization stems from reduction/ compression of consumption in household sectors due to restriction on cash withdrawal. In this context, it is important to mention that the RBI has revised the growth forecast from 7.6% to 7.1% for 2016–17. There are other estimates indicating downward growth of 6.8%. The Reserve bank views that there will be short run disruption in the economic activity but the impact will "ebb with progressive increase in the circulation of new currency notes and greater usage of non-cash based payment instruments in the economy" [10]. The one-time cost of implementing demonetization also affects economic growth in the current year. According to Mahesh Vyas, MD & Chief Economist, The Centre for Monitoring Indian Economy, this cost calculated for 50 days, amounts to 40% of annual income tax revenue [11]. Further, if the rupee depreciates, the economic growth in India could suffer.

3.3 Inflation

Interestingly, the impact of demonetization on inflation is ambivalent. Only after obtaining the data of Q4 - 2016–17, one will know the exact situation. The perishable commodities were off loaded at very low prices or discarded, as there were no buyers in the market. Consequently, there could be some disinflation in items like vegetables and fruit. As per projections, the stickiness in the non-food and non-fuel inflation could balance the impact. Overall, there is a concern, as mentioned by RBI, to maintain the inflation rate at 5% in Q4 of 2016–17.

3.4 Black Money and Counterfeit Currency

Black money in India, as per the World Bank report 2012, is one fifth of the GDP of the country. Such a large magnitude creates inequality. It also restricts the revenue to the government and impacts government spending for common good. The objective behind demonetization was to contain black money. In terms of magnitude, the withdrawn currencies have come back to the banking system. Black money is not necessarily money. It is money in different pockets and forms instead of being with the government: "it can be in different forms other than cash" [12]. With regard to counterfeit currency, the impact is two-fold: firstly, seizure of counterfeit bills in lieu of discontinued bills of ₹ 500 & ₹ 1000; secondly, enhanced security features in the new bills would make it difficult to counterfeit the new bills. How soon the criminals eventually catch up, remains a guess.

3.5 Impact on the Union Budget

Initially, the government had estimated that substantial amount of demonetized currency will not come back to the system. This currency will seize to be the monetary liability of the RBI, recorded as non-monetary liability. Hence, the reserves of RBI will increase in its balance sheet, thereby making a space for higher profit transfers to the government. Although the RBI has not clarified its stance on this matter, it would be prudent to suggest that this is an erroneous view.

It is also held that the a part of the currency deposited with the banks could be black money and if it comes under the amnesty scheme announced by the government (50% Tax and 25% investment in development bonds), there would be some revenue gain for the government. The relevant data are not available in public domain to prove this point.

An analysis of the liquidity management process (through the cash management bills) undertaken by the RBI to address the deposit of the bills in the banking system, revealed the adverse impact on the budget, which is the interest outgo from the government budget.

4 Impact on the Service Sector

At an aggregate level, the service sector has taken a hit. Construction, hotel and other service sectors where cash transactions are high, have shown a massive slow down. In order to unfold the micro level impact, we engaged in a natural experiment study with common-man in Mumbai city. This is a city of large number of migrant workers in service sector. Mumbai also has high population density [20980/sq. km] that aggravates a rupture in services, lending it amenable to survey. Mumbai residents possess mobile phones and have access to TV; they are well versed in banking transaction and consequently more aware of the social and financial environment. We approached people at the market place and the local railway station in Vile Parle area (Fig. 1) and received 170 responses. This comprised people earning between ₹3000–10,000 per month. This study represents the situations in most metropolitan towns in India. The narrative in remote villages would be different and the impact of demonetization not as severe.

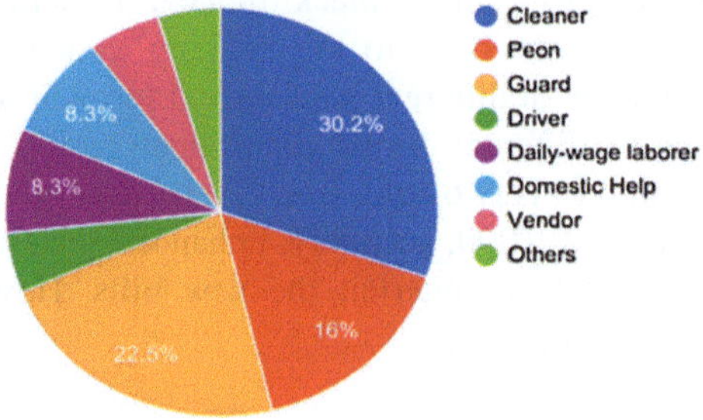

Fig. 1. Profession of the respondents

After analyzing demonetization and assessing the situation at the macro level, we designed a questionnaire with following hypothesis:

- Low income service providers are not aware of the concept or the consequence of demonetization.
- The man on the road is not aware of black money but has a perception that rich people have black money.
- Demonetization is pro-poor because it is anti-rich.
- The role of RBI as the creator and destroyer of money did not surface since the Prime Minister became the visible face of demonetization.
- The ignorance of the masses helped the government surmount the procedural ambiguity and opacity.

4.1 Low Income Service Providers are Not Aware of the Concept or the Consequence of Demonetization

Around 71.2% of our respondents came to know about demonetization from T.V News. The information dissemination on popular media channels (print as well as T.V), did not elaborate reasons for demonetization amidst the discussions lasting for 2

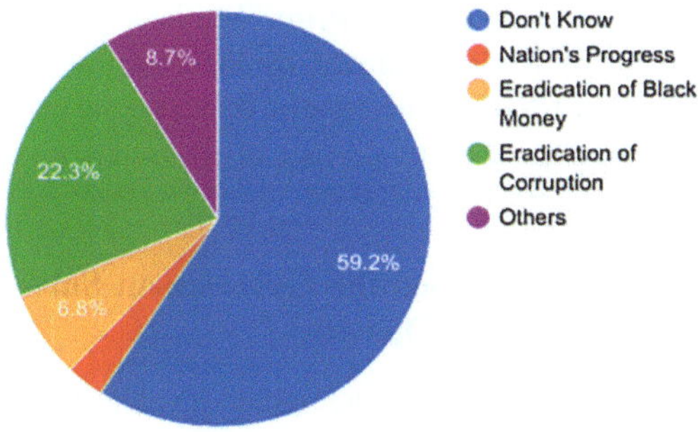

Fig. 2. Rationale for demonetization

months, resulting in lack of insight and lack of awareness. This is evident in our survey where majority of the respondents "Don't Know" (Fig. 2). Thus, our first hypothesis "Low income service providers are not aware of the concept or the consequence of demonetization" is vindicated.

4.2 The Man on the Road is Not Aware of Black Money but Has a Perception that Rich People Have Black Money

Our survey contained two questions namely, "Do you know what black money is?" and "Do rich people have black money?" Nearly 50% of the respondents were not aware of "What is Black Money?" However, 73% of the respondents have a perception that "rich people have black money". Thus, our second hypothesis "The man on the road is not aware of black money but has a perception that rich people have black money" stands vindicated.

4.3 Demonetization is Pro-poor Because It is Anti-rich

This follows from the second hypothesis. 60% of our respondents were "Happy with Demonetization". The reasons for this happiness were that "the rich will be penalized" for 67.9% of the respondents and "something better will happen" for 48% of the respondents. This societal perception is atypical of the class structure rampant in India. Anecdotal evidence revealed that the move of the government had made the common man happy because for once, the rich and poor were brought on the same level. Thus, our third hypothesis "Demonetization is seen as pro poor because it is anti-rich" stands vindicated.

4.4 The Role of RBI as the Creator & Destroyer of Money Never Surfaced Since the Prime Minister Became the Visible Face of Demonetization

According to the RBI Act 1934, RBI is the sole creator and destroyer of money. However, the masses do not know this fact. Therefore, hearing the Prime Minister's speeches, the respondents inferred that the Prime Minister is the sole authority for the demonetization process. This is corroborated by 90% of our respondents. Thus, our fourth hypothesis "The role of RBI as the creator & destroyer of money never surfaced since the Prime Minister became the visible face of demonetization" stands vindicated.

4.5 Ignorance of the Masses Helped the Government Surmount the Procedural Ambiguity & Opacity

The common man was exploited by both the authorities and the employers. Anecdotal evidence suggestes that there are cases of postponement of marriages, hardships in receiving medical care, loss of job, particularly for daily laborers in small service sector, etc. Majority of our respondents faced inconvenience for over a month and approx. 80% of them stood in the queue for more than 2 h (Fig. 3), without raising their voice, because of their perception that "demonetization is a penalty for the rich". Moreover, 92% of the respondents have stood in the queue for others. This is a clear case of exploitation of the service providers by the employer. Thus, our fifth hypothesis "Ignorance of the masses helped surmount the procedural ambiguity & opacity" is vindicated.

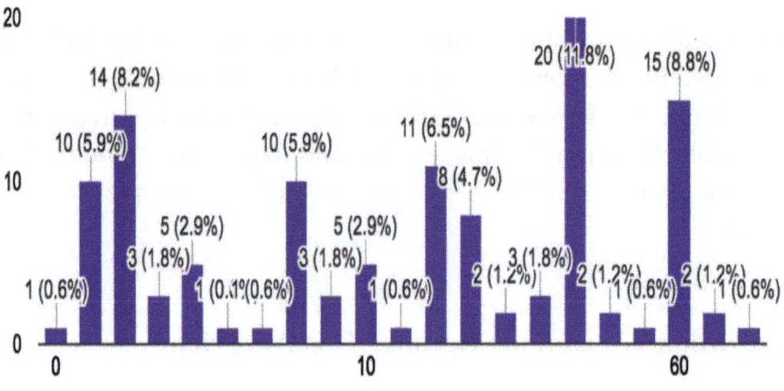

Fig. 3. Number of days that the respondents suffered due to demonetization. The X axis denotes the number of respondents and the Y axis indicates the number of days (*0 to 90*)

5 Conclusion

Contrary to the view of the authorities that demonetization was implemented with adequate planning, our paper infers that the process was carried out without adequate technological and administrative preparedness both, by the government and the RBI. There has been ambiguity and non-transparency with regard to the role of the RBI in the whole process. The authorities could get away with harsh measures because of the class structure of the Indian economy and managed the sentiment of the common people. The objective of the exercise kept on changing from time-to-time: from fake currency and black money to terrorist financing and to digitization and cash-less economy. Our natural experiment suggests that it is only due to the ignorance and resilience of the common people that the implementation of demonetization did not result in any social upheaval. Hardship is not unknown to a common man in India. In fact, the idea of personal suffering is considered noble to build one's 'karma'. When government insisted on opening 23.3 million *Jan Dhan* bank accounts, the common folk complied with the order. While middle class, familiar with technology, activated digital transactions, the anecdotal evidence suggests that the first time account holders did not know how to operate their account. Eventually, this might result as a move towards greater financial inclusion. As for time and income loss, the narrative in metropolitan towns is different from villages in India where people generally know each other well enough and where differed payments are usual way of existence.

In our considered view, to address the fake currency issues, the RBI needs to focus on strengthening its currency management architecture; and to cleanse the black money from the system, the government needs to focus on strengthening the efficiency of tax machinery.

Acknowledgements. Authors acknowledge the stimulation by way of discussions and analysis provided by Ms. Rujuta Karkhanis & Mr. Mohd. Adib Mustafa, students of MBA at NMIMS, Mumbai. Mr. Ashok, a member of administrative staff and Ms. Rajashree helped the authors speak to Marathi respondents.

References

1. Kohli, R., Saharay, A.: Journal of Income & Wealth, July–December 2011
2. Report of the Central Board of Directors on the working of the Reserve Bank of India for the year ended June 30, 2016 submitted to the Central Government in terms of Section 53(2) of the Reserve Bank of India Act (1934). https://rbidocs.rbi.org.in/rdocs/AnnualReport/PDFs/0RBIAR2016CD93589EC2C4467793892C79FD05555D.PDF
3. Demonetisation: To Deify or Demonize? http://indiabudget.gov.in/es2016-17/echap03.pdf
4. Only 16 of every 250 fake notes in India are detected by authorities. http://www.hindustantimes.com/india-news/only-16-of-every-250-fakes-notes-in-india-are-detected-by-authorities/story-i3uzdSZU8mj66dFumGW5jM.html
5. 20 days after demonetisation, 3.4% of all notes returned are counterfeit. http://timesofindia.indiatimes.com/india/20-days-after-demonetisation-3-4-of-all-notes-returned-are-counterfeit/articleshow/55800888.cms
6. Demonetisation to have positive impact on economy: Arun Jaitley. http://economictimes.indiatimes.com/news/politics-and-nation/demonetisation-to-have-positive-impact-on-economy-arun-jaitley/articleshow/55601371.cms
7. Demonetisation: Manmohan Singh's full speech in Rajya Sabha. http://indianexpress.com/article/india/india-news-india/demonetisation-manmohan-singhs-full-speech-in-rajya-sabha-4392829/
8. RBI facing 'reputational' risk, says former governor YV Reddy. http://www.livemint.com/Politics/7s15KPbTeizn1i5ZBKI5VL/RBI-facing-reputational-risk-says-former-governor-V-Y-R.html
9. Interview to Karan Thapar on India Today Television, 10 January 2017

10. RBI cuts growth forecast to 7.1%, sees short-term disruption. http://www.newindianexpress.com/business/2016/dec/07/rbi-cuts-growth-forecast-to-71-sees-short-term-disruption-1546595.html
11. Transaction cost of demonetisation estimated at Rs.1.28 trillion. https://www.cmie.com/kommon/bin/sr.php?kall=warticle&dt=2016-11-21%2015:12:31&msec=360
12. Black Money Antidote Stifles Economy. http://epaperbeta.timesofindia.com/Article.aspx?eid=31808&articlexml=Black-Money-Antidote-Stifles-Economy-08122016017007

The Key Success Factors in the Development of Platform-Based Business in China - Case Study on Alibaba and Jingdong

Wenjuan Zhang(✉), Ming Chen, Wenjia An, and Jianan Zhang

Sino-German Applied Science College,
Tongji University, Shanghai 200092, China
{zhangwenjuan, chen.ming}@tongji.edu.cn

Abstract. Alibaba group and Jingdong group are two outstanding representatives of e-commerce providers in China. But their business modes are completely different. And the outside evaluations are sometimes completely different. After theory research, this article believes that the future of B2C platform business is user-centric, digital-driven, omni-channel service platform. Then by using comparison method and case study, we concludes that the most important factors for a successful E-business service platform in China is not technology, but high-qualified products and services and employees' welfare and benefits.

Keywords: Business model · Service platform · User-centered

1 Preface

The new industrial revolution is a profound change which is triggered by technology, will ultimately change our entire human society, lead us enter into an era of knowledge and creativity, the construction and advancement of social cultural environment is the necessary precondition. In this sustainable and comprehensive process, human being will become the most important manufacturing factor beyond land and capital, and play a more essential role in the industrial production. "Take people as the foremost" is the core principle of this revolution. E. L. Trist, the founder of sociotechnical system, and his colleagues in the UK Tavistock Institute insist on: in a sociotechnical system, manager's main task is to ensure the coordination of social system and technical system. Technical system is the intermediary between organization and its environment. But when the conflict happens technical system usually should make some changes to adapt to the social system [1].

2 The Situation of E-commerce Retail Trade in China

E-commerce has revolutionized the retail trade. E-commerce industry in China develops ahead of the world, and has become a major engine of this decade's rapid economic growth in China. It is said that in the next five years, China's e-commerce

T. Ahram and W. Karwowski (eds.), *Advances in Human Factors, Software, and Systems Engineering*, Advances in Intelligent Systems and Computing 598,
DOI 10.1007/978-3-319-60011-6_19

will continue an average of 29% compound growth rate, greater than those of Europe and the United States and the other advanced countries. China is also the largest e-commerce market in the world by value of sales, with an estimated US $899 billion in 2016. As the main component of the service economy, e-commerce will face more challenge in the future and keep on developing towards more platformed, socialization and intelligent model.

Chinese e-commerce enterprise Alibaba (hereinafter abbreviated as Ali) is ranked the world's second most valuable retail brand with a brand value of 49.30 billion USD [2]. Since 2009, Ali began promoting Nov. 11, a date made up of four 1 s, as a shopping festival. And this date now turns out to be the country's busiest day for online retailers. On that single day of 2016, the total turnover reached 17.8 billion USD, wireless transactions accounted for more than 81.87%, covers 235 countries and territories [3]. However, Ali's closing stock price on Wall Street is down 2.41% from the previous day. And actually not only 2016's double eleven, for at least three consecutive years the declining trends are the same since 2014 (see Fig. 1).

Fig. 1. The price of Ali's stock in the consecutive 3 year after 11/11 [4]

Jingdong (hereinafter abbreviated as JD) was founded in July 1998. Started with a focus on online direct sales of electronics products, soon accepted other retailers as an e-commerce marketplace [4]. JD delivers a majority of the orders directly to customers. Recently, the US retail industry magazine Stores & Deloitte announced the 2017 global retailer list of 100. It shows that JD is one of China's two platform company listed, and JD also ranks second among the world's top ten fastest growing retailers list [5]. JD shares the very similar development process and business philosophy with Amazon, which has become the world 's largest B2C e-commerce company. The two companies' annual revenue were exponential growth since their setup over the years. But Amazon has turnaround in 2015 after 20 years' continuous loss, JD is still in a loss state, and the trend is expanding.

What's the difference between Ali's and JD's business model? Why Ali's "bumper harvest" of online transactions is not accepted by the capital? Why JD is continuing loss but accepted widely by stakeholders? This article tries to find out some reasons behind these phenomena and hope can get some factors to help Chinese platform companies develop continuously and efficiently.

3 The Development Trend of Retail E-commerce

3.1 Omni-Channel Business Model

Nowadays, consumers are increasingly inclined to buy personalized products and services and they also don't want to wait. The traditional distribution economy is gradually replaced by different business mode such as cross-channel, omni-channel, social, ambient, everywhere and no-line commerce concepts. Many researchers believe omni-channel business model will become the next generation of retail. In this model, online and off-line channels are blended into a single 24/7 shopping experience. Its a cross-company digital value chain including a service platform to bring up a new generation of logistics companies with primary focus on the orchestration of global supply-chain networks that integrate swarms of production enterprises and logistics providers. The first phenomenon of this model is that traditional brick-and-mortar commerce begins to open e-shops in parallel to their physical store or cooperate one or more e-commerce platform such as ebay, Amazon, and Meinpaket. de. On the other hand, on-line platform traders are also trying to be involved the traditional channel or physical infrastructure.

Omni-channel business should match different standard networks and established assets efficiently, tailor to meet the needs of each single channel,provide individually configured solutions. Many neutral intermediaries who actually without physical assets of its own, but can balance the interests of providers and customers, supplant manufacturers and service providers by gaining sovereignty over the data. Such as 4PL providers, special expertise, special service provider, many local and small logistics providers will benefit from this model. Some famous retail provider such as Amazon, ebay, Ali are working on vertical integration to establish their own end-to-end global omni-channel supply networks [7].

3.2 Some Features of Omni-Channel Business Model

It's a Platform-Based Ecosystem. Omni-channel business act as a service platform to make different companies or partners setup seamless, largely automated and legally compliant collaboration. Whoever controls the service platform will can gain control of the entire value chain [8]. According to an survey taken by Accenture company, about 81% of executives said platformed business model will be core to their growth strategy within 3 years [9]. Platform businesses will control the access to markets and customers services are decisive for user experience and success. Platform is a cross-company value network which often has little or no value of its own. The value of a platform reflects in its ability to provides infrastructure and tools on which others can build or add value.

The platform can fully automatically match requests and supplies of products and services, can anticipate and prevent the potential disruption or goal conflicts at all times. Thus can increase prediction accuracy, fast delivery through efficient network planning, fulfill real-time information track and sharing and reduce chain risk through predictive risk evaluation ability. Uber, Airbnb, Ali are good examples of such platform enterprises. It is believed that service quality, user experience, automation degree, cost reduction, time and effort reduction, safety are the key factors for the success of a platform [10]. Generally, platform can be evaluated according to cost-effectiveness, user acceptance, IT security and data protection reliability from the usage perspective of the consumer [8].

It's a Data-Driven Network. Data is looked as the most important material of this century and the value-added control points of a company. Data technology can combine and analyze the data taken from hereogeneous physical systems and services to give smart suggestions, is able to anticipate customers' requirements accurately, can support predictive algorithms to improving the quality of business process forecasts, can boost process efficiency and service quality by speeding up delivery time and enhancing capacity and network utilization. Data carries huge untapped potential for optimizing capacity utilization, reducing risk, improving customer experience, and creating new business models.

It Centres Around the Users. In one of its research report, Accenture found out 83% of the executives believed the digital economy is driving a major shift in power from the supply-side economies of scale to demand-side economies of scale based on the network effects of two-sided markets, where value is created in platform ecosystems [9]. Only when the users employ services in their respective roles as consumers, employees, citizens, patients and tourists, does the platform can get value of their own. customers. As far as the customer is concerned, smart services mean that they can expect to obtain the right combination of products and services to meet the needs of their current situation, anytime, anywhere (see Fig. 2) Smart service providers therefore require an in-depth understanding of their users' preferences and needs.

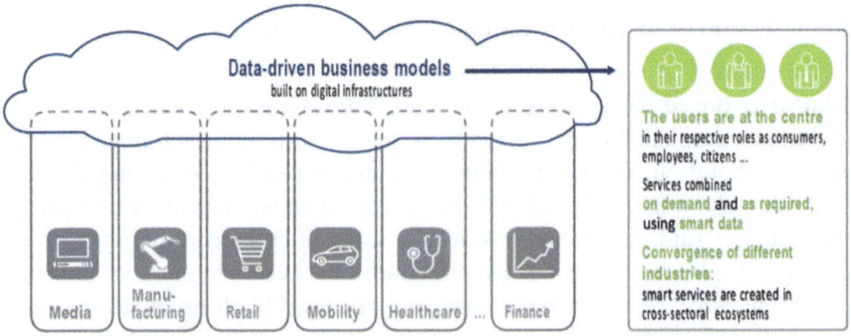

Fig. 2. Smart services: the user is at the centre [8]

3.3 Comparison Between Ali and JD as a Service Platform

Ali and JD are two outstanding retail e-business enterprises. Both of them are walking towards more successful transaction service platform. Hereby we want to analysis them as follows.

Customer Service. With the economic development, public consumption habits have changed. In October 2016, "Forbes" published an article which points out: the price is no longer the most concerned factors, but the product quality and service will be payed more attention.

Since its setup, JD has insisted on the business philosophy of "customer first". Its founder repeatedly stresses that "customers always look forward to the quality, service and price. From the data of 2014 to 2016, given by China Electronic Commerce Research Center, [11] we can find out that JD has a high response rate of customers' complaints, while the complaint ratio has kept very low and its customer satisfaction ratio is high. However Ali still hasn't found a good solution to monitor products quality sold on its platform, and hasn't got a good solution to deal with various buyers' complaints effectively. Both Ali's feedback rate and feedback speed are disappointing. Its feedback rate for three consecutive years are all less than 3% (see Fig. 3). Its customer satisfaction rate for three consecutive years ranked No. 20th, far behind JD who has been the top three at the same time period.

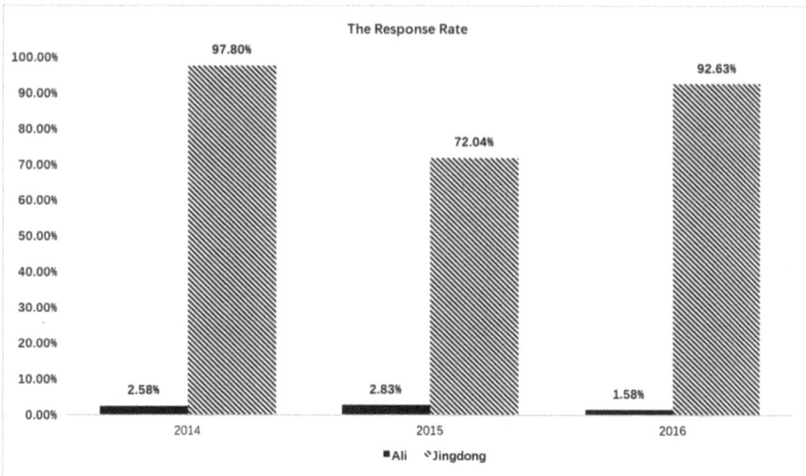

Fig. 3. The comparison of response rate between Ali and JD

In order to achieve faster customer response speed, JD independently develops a set of ERP system, and cooperates with other enterprise resource software vendors to improve their information processing speed greatly through the whole business process. Ali is also working with Yongyou, an Asia-Pacific large-scale enterprise management software, Internet services and financial services provider, by linking Yongyou's network with the Ali's ecological resources, puts the e-commerce and cloud computing services into business management and operational services, therefore it can create a polymerization effect and new value.

Continuous Enlarge the Application of Information Technology. In 2014, JD established cooperation relationship with Tencent, one of China's largest internet integrated service provider, ask the latter to provide JD WeChat's and mobile QQ's direct entrance and other major digital platform support. And they also cooperated in diversifying online payment channels.

Since 2004, Ali started to build Alipay platform. In the second quarter of 2014, Alipay became the world's largest mobile payment platform. Alipay combines the payment, life services, government services, social, financial, insurance, public welfare and other scenes and industries. It provides customers with more convenient services. Currently Alipay has become the life service platform owning more than 450 million real-name users.

Setup of Cross-Company Network. As a B2C electric business, in order to ensure the quality of products, JD purchases products in three ways: manufacturers direct supply, brand settled and procurement. And under the principle of win-win, JD opens its owned warehouses to satisfy other logistics companies' requirements.

In 2016, Ali ranked first in China's B2B service providers, the market share rose 0.5% and reached to 46.20% [12]. On its platform, Ali provide a considerable number of value-added services to its settled vendors and manufacturers. For example, Ali's foreign trade partners can participate in O2O line exhibitions, obtain finance support, credit protection and other relevant information at a lower cost.

Create More Job Opportunity. According to JD's "2016 corporate Citizenship Report", JD has created directly and indirectly more than 4.1 million jobs [13]. And correspondingly, in recent years, Ali retail business ecology has created more than 15 million jobs, drive about 30 million indirect employment. It is estimated that Taobao and Tianmao, two key B2C online companies under its name, through stimulating new domestic demand, have brought the upstream manufacturing tax revenue growth of nearly 180 billion RMB [14].

Initially Form Digital Service Platform. JD is also entering the field of cloud computing. With the application of the world's leading cloud computing technology, JD cloud can provide cloud computing integrated service, can calmly resist the impact of a large number of traffic bursts, Can complete 100% of the order in the cloud. During the promotion period, JD cloud bandwidth expansion of hundreds of G [15].

Ali's business is not limited to e-business, but also includes cloud computing, mobile Internet and others. Its financial results of the second quarter of fiscal year 2016 shows that Ali cloud computing, Internet infrastructure services continue to maintain rapid growth, and its quarterly revenue reaches 0.15 billion USD, with increase of 175%, the growth rate over the previous quarter 126%. The number of cloud computing's paid users has reached more than 0.5 million. Following the Amazon AWS and Microsoft Azure, Ali has become the world's third largest cloud computing service providers [16].

The Comparison of these Two Companies as Service Platform. According to the above description, we can get the comparison of these two companies as a service platform as follows (Table 1).

Table 1. Compare the features of Ali and JD as a service platform

Feature of smart services [8]		Features of Ali	Features of JD
1	User-central	not so good	better
2	Cross-company and cross-sector	good	good
3	Data-driven	good	good
4	Extremely agile–short release cycles	good	good
5	Data and algorithms increase value–added–service	good	good
6	Lateral business benefits	good	good

Remarks: [a] the score we gave here is a based on the relative comparison between two companies, not an absolute one. And mainly according to the information we described above and for the purpose of research.

4 The Key Factors in Shift to Platform-Based Business in China

From the above statement and comparison table we can find out that both Ali and JD are walking towards a good service platform providers. Of course the degree of these features they satisfied is different. But the largest difference lies in the first one: "user-central". And we also can state that the future economy is digital-based, but the main success factor of data economy cannot only be locked on technology, and firstly should be customer-central. Ali is doing better towards a DT company, but JD focus more on the customer. We think that's the main reason to explain the different choice from the stockholders for these two companies. In Germany, it is said that successful new business models will only emerge where complex smart products and smart services are combined and orchestrated by well-trained employees, or smart talents [8]. But that conclusion is based on the needless-to-worry qualified German product and their reliable corporate reputation mechanism. In China, we believe a successful B2C platform provider should firstly provide the customers high-qualified products and services, and at the same time focus on their employees' welfare and benefits.

4.1 Quality of Product

From it's foundation, JD keeps on putting customer at the center, keeps on pursuing high product quality, sets higher seller's entry threshold. Although JD also faces with some unqualified goods complaints, with 60% of the goods directly purchased from manufacturers, it can give its customers a big promise of product quality. To strengthen supervision of trading products, in April 2016, JD established cooperation relationship with state administration bureau of quality supervision, one of China's government department who supervises and inspects product quality. Thus its consumers can easily check the quality of purchased products on the government's commitment information platform. And from February 16, 2017, JD also authorizes some professional third-party institutions to check the product quality.

On the contrary, although Ali provides retailers a very good platform and additional services, it can't give buyers a reliable product quality guarantee. Its buyers keep on complaining product quality, price fraud and fake sales and related problems. According to "2016 network transactions commodity quality special sampling results", taken by the general administration of Industry and Commerce, Tmail, which Affiliates to Ali, has the most large number of unqualified products, and Ali's another affiliated enterprises named Taobao also lists the forefront. Ali has realized their Product quality dilemma and has taken some measures to strengthen supervision of product quality. for example, the usage of self-test method, which lets their employees to buy some "high-risk" goods for inspection. However, because Ali's business philosophy focuses more on the seller's side and there are too many sellers on its platform to be examined, the effect is still very limited.

4.2 Quality of Service

The concept price is put in the second place among JD's "quality, service and price"operating principle. Thus we can find out the importance of service for JD. For an e-business trade platform, the most important service is stable and timely logistics services. JD relies on self-built logistics infrastructure while Ali mainly relies on the third-party logistics. Self-built logistics is characterized by large early investment, but after that it can achieve economy of scale and reduce logistics costs and improve transport speed. JD layouts logistics facility firstly from a point then to an area. It builds logistics centers firstly in the first-tier cities which act as the transfer center to store and transport goods to the surrounding second-tier cities. These first-tier cities are characterized by densely populated, with obvious geographical advantages. They currently owns 118 large warehouses in 79 cities with a total area of about 2.3 million square meters around the country. In recent years, JD cherishes the ambitious vision to build "Asia on the 1st" project, and continuously improve intelligent, automated level of its logistics service.

Recently Ali realizes the importance of logistic. But it focuses on the construction of virtual logistics platform named "Cainiao". After 2015, Cainiao began to develop a principle said:" build warehouses, but do not operate them; do logistics technical support, but do not involve in the direct distribution operation".

4.3 The Employee

Employee is the very important and active factor for a company's steady operation. This idea can be found from JD's human resource management principle. The salary of JD's couriers is very high compared with their peers. From the report of JD 2016 annual employee benefits, we find out JD spent more than 2.7 billion RMB on its employees. And Ali also launched a 3.54 billion welfare program to benefit employees, added another 3 billion RMB interest-free home loans, 500 million RMB education fund, 40 million RMB price subsidies. And the employees, who has worked at Ali for more than 3 years, can get about 0.2 or 0.3 million RMB interest-free loans to help them buy apartments or subsidize other huge expenses [16]. Overall, the two companies all do very well in their individual way. So their employees are more active and motivated than the other peers. That one reason to help the two companies develop fast and successfully.

5 Conclusion

The future of e-business retail trade business mode is a kind of omni-channel platform economy, which is a data-driven user-focused service platform. The success factors in this trend are different according to individual country's condition. After the detailed data analysis and case study, we agree with the statement that the mission of an enterprise should be to provide a better proposition for users, contributes to social welfare and permits workers to enjoy decent working conditions [8]. And we think the key success factors in the development of platform-based business in China should be quality of products, quality of services, and the focus on their employee.

References

1. Sociotechnical system-Wikipedia. https://en.wikipedia.org/wiki/Sociotechnical_system
2. Top 25 Most Valuable Global Retail Brands 2016/2017 - China Internet Watch. https://www.chinainternetwatch.com/19579/top-global-retail-brands/
3. E-commerce sales reach $17.8 billion on Singles' Day. https://www.internetretailer.com/2016/11/11
4. Alibaba (NYSE: BABA). https://xueqiu.com/S/BABA
5. JD.com (JD) Intro, Stats, News & Insights - China Internet Watch. https://www.chinainternetwatch.com/tag/JD/
6. Global Top 100 Retailers Rankings. http://mt.sohu.com/20170122/n479314128.shtml
7. DHL:Logistics Trend Radar:Delivering insight Today, Creating Value Tomorrow (2014)
8. German National Academy of Science and Engineering: Smart Service Welt: Recommendations for the Strategic Initiative Web-based Services for Businesses, final report (2015)
9. Accenture: platform economy: technology-driven business model innovation from outside in (2015)
10. Walter Mattauch: Economic relevance of smart services technologies. http://www.innovation4.cn/library/r1395
11. China Electronic Commerce Research Center. http://www.100ec.cn/
12. The first half of the B2B market share ranking: Ali first seven core. http://b2b.toocle.com/detail–6358291.html
13. Jingdong released corporate citizenship report. http://mt.sohu.com/
14. Alibaba: New Trends and New Explorations of China's B2B Trends in 2016. http://www.imxdata.com/archives/1530
15. Baidu Encyclopedia - Jingdong cloud. http://baike.baidu.com/
16. The Growth of Cainiao. https://www.huxiu.com/article/152077.html

Author Index

T. Ahram and W. Karwowski (eds.), *Advances in Human Factors, Software, and Systems Engineering*, Advances in Intelligent Systems and Computing 598,
DOI 10.1007/978-3-319-60011-6

MIX
Papier aus verantwortungsvollen Quellen
Paper from responsible sources
FSC® C105338

If you have any concerns about our products,
you can contact us on
ProductSafety@springernature.com

In case Publisher is established outside the EU,
the EU authorized representative is:
Springer Nature Customer Service Center GmbH
Europaplatz 3, 69115 Heidelberg, Germany

Printed by Libri Plureos GmbH
in Hamburg, Germany

MIX
Papier aus verantwortungsvollen Quellen
Paper from responsible sources
FSC® C105338

If you have any concerns about our products,
you can contact us on
ProductSafety@springernature.com

In case Publisher is established outside the EU,
the EU authorized representative is:
Springer Nature Customer Service Center GmbH
Europaplatz 3, 69115 Heidelberg, Germany

Printed by Libri Plureos GmbH
in Hamburg, Germany